Laws of Nature and Chances

Breathing Fire into the Equations.

Laws of Nature and Chances

What Breathes Fire into the Equations?

BARRY LOEWER

UNIVERSITY PRESS

Great Clarendon Street, Oxford, OX2 6DP,
United Kingdom

Oxford University Press is a department of the University of Oxford.
It furthers the University's objective of excellence in research, scholarship,
and education by publishing worldwide. Oxford is a registered trade mark of
Oxford University Press in the UK and in certain other countries

© Barry Loewer 2024

The moral rights of the author have been asserted

All rights reserved. No part of this publication may be reproduced, stored in
a retrieval system, or transmitted, in any form or by any means, without the
prior permission in writing of Oxford University Press, or as expressly permitted
by law, by licence or under terms agreed with the appropriate reprographics
rights organization. Enquiries concerning reproduction outside the scope of the
above should be sent to the Rights Department, Oxford University Press, at the
address above

You must not circulate this work in any other form
and you must impose this same condition on any acquirer

Published in the United States of America by Oxford University Press
198 Madison Avenue, New York, NY 10016, United States of America

British Library Cataloguing in Publication Data

Data available

Library of Congress Control Number: 2024934387

ISBN 978–0–19–890769–5

DOI: 10.1093/oso/9780198907695.001.0001

Printed and bound in the UK by
Clays Ltd, Elcograf S.p.A.

Links to third party websites are provided by Oxford in good faith and
for information only. Oxford disclaims any responsibility for the materials
contained in any third party website referenced in this work.

Contents

Preface	vii
1. Introduction	1
2. Non-Humean Accounts of the Metaphysics of Laws	7
3. The BSA Humean Account of the Metaphysics of Laws	18
4. Objections to the BSA	29
5. Problems with Perfectly Natural Properties	53
6. Super-Humeanism	71
7. The PDA	79
8. The PDA and Chance	96
9. Special Science Laws and the PDA	116
10. Realism, Relativism, and Reference	124
11. Reprise and Conclusion	132
References	141
Index	147

Preface

Stephen Hawking famously asked, "What is it that breathes fire into the equations and makes a universe for them to describe?" This book is my attempt to answer Hawking's question. I understand him to be asking what distinguishes lawful from merely true propositions or equations and what determines the fundamental ontology. Hawking may also be asking why the universe exists at all. My answer says something about what the universe must be like for there to be laws and a fundamental ontology but why there is a universe at all is a mystery beyond the scope of this book.

This book is the culmination of many years of thinking about laws and chances. It is based on but goes far beyond numerous papers I have published since 1996 on laws, probabilities, and counterfactuals. In it I develop a novel account of laws, chances, and fundamental ontology based on David Lewis's Best Systems Account but radically different from his in not being committed to Humean metaphysics. In formulating my account, I take lessons from Lewis's teacher W. V. O. Quine concerning fundamental ontology. The answer I give to Hawking's question, likely one Lewis would not have accepted, is that it is the activity and practice of science that breathes fire into the equations and determines what makes fundamental ontology.

I am enormously indebted to many friends and colleagues for comments and criticisms.[1] Most importantly without the support of David Albert and Jenann Ismael this book would not have existed. Without the support and love of Katalin Balog and Marjorie Loewer I would not

---

[1] Among my friends and colleagues in addition to David and Jenann who gave me helpful comments, criticism, and sometimes objections are Marshal Abrams, Diego Arana, Marvin Belzer, Harjit Bhogal, Thomas Blanchard, David Builes, Eddy Chen, Heather Demarest, Alison Fernandes, Veronica Gomez, Bixin Guo, Mike Hicks, Carl Hoefer, Jake Khawaja, Mark Lange, Dustin Lazarovici, Ernest Lepore, Milan Loewer, Tim Maudlin, Vishnya Maudlin, Jill North, Walter Ott, David Papineau, Georges Rey, Howard Robinson, Ted Sider, Russell (Philip) Stetson, Michael Strevens, and Isaac Wilhelm.

have continued to exist to write it. Katalin Balog went over the entire manuscript improving it enormously.

I am also grateful to the Templeton Foundation for awarding me a grant that allowed me time off from teaching to finish the book.

1
Introduction

The idea that the universe is governed by laws of nature has precursors from ancient times and in non-European cultures, but the view that one of the main aims of the sciences is to discover fundamental universal mathematical laws only became established during the sixteenth and seventeenth century in Europe, when it replaced the then prevalent Aristotelian conception of science.[1] The most prominent promoters and developers of the new science were Galileo, Descartes, and Newton. Descartes, inspired by Galileo, dreamed of an elegant mathematical theory specifying laws that describe the motions of all material objects, and Newton in his *Principia* went a long way towards making this dream come true.

Historians of science mostly agree that the emergence and development of the idea of a law-governed universe in the seventeenth century were strongly influenced by theology, and especially by the views that matter is passive, that God is the source of all motion, and that these motions can be described by mathematics.[2] This contrasted with the Aristotelian account of nature according to which the universe consists of diverse kinds of entities, each with their own powers, goals, and activities. The paradigm science for Aristotle was biology, which studies living beings whose natures are to move each in their own ways for their own ends. By contrast, on the law-governed view of the world, the

---

[1] The idea that the universe is subject to natural laws is sometimes said to have originated with Anaximander and other pre-Socratics. But they did not propose mathematical formulas to describe the motion of objects and did not begin a search for laws in the way that Descartes and his contemporaries proposed, nor did they have the contemporary view that proposals should be tested by observation and experiment.
[2] See Peter Harrison (2008) and John Milton (1988). A dissenter from this view is Jane Ruby (1986) who downplays the role of theology in the development of the concept of laws. Of course, Descartes and some of his contemporaries thought that God allows human minds to also move matter and so the laws he imagined could be "broken" in certain ways by minds.

Laws of Nature and Chances: What Breathes Fire into the Equations? Barry Loewer, Oxford University Press.
© Barry Loewer 2024. DOI: 10.1093/oso/9780198907695.003.0001

motions of all material objects are due entirely to the uniform operation of mathematical laws that carry out God's will. The historian of science Peter Harrison writes:

> Unlike the ontologically rich Aristotelian world, the sparse world of atoms or corpuscles was unpopulated by the qualities, virtues, active principles, and substantial forms that had once invested nature with significant causal agency. This was a causally vacant cosmos that would be receptive to the direct volitions of the Deity.[3]

Samuel Clarke, Newton's spokesperson in his debate with Leibniz, called laws "God's volitions."[4] They are the source of all motion and change.[5] Descartes believed that laws could be expressed as theorems of a system characterized by a few mathematical axioms that covered the motions of all matter. In his *Principia* Newton showed how this might be done. Laws were thought to reflect God's nature. They are simple, mathematical, eternal, universal, exceptionless, and deterministic. On this view laws both systematize and govern the world in accord with God's plan. That humans can discover them was due to divine benevolence. Descartes dreamed that a new mathematical science would enable their discovery.[6]

Without Newton, Descartes's dream might have died. Instead, it initiated a program that led to Lagrangian and Hamiltonian classical mechanics, Maxwell's electromagnetic theory, thermodynamics, statistical mechanics, the atomic theory of matter, molecular chemistry, special and general relativity, quantum field theory, and in the twentieth century evolved into Steven Weinberg's dream of a final theory or theory of everything (TOE); a theory that specifies the universe's fundamental

---

[3] Harrison (2008) 9.
[4] "The Course of Nature, cannot possibly be anything else, but the Arbitrary Will and pleasure of God exerting itself and acting upon Matter continually"; Clarke (1738) 150.
[5] But Newton didn't think that matter is passive. His view was that matter has the power to obey laws and so powers and laws are interdependent; see Psillos (2014) for discussion.
[6] Important to the development of this view—realized in Newtonian mechanics—was the discovery of simple principles that describe the motions of planets, and the idea proposed by Galileo and others that the principles that describe the motion of celestial and terrestrial object are the same or closely related.

ontology and a system of laws that by unifying general relativity and quantum field theory completely covers all natural phenomena.

The laws of Newtonian mechanics are universal and deterministic. This set the stage for diminishing the role of God in science. If the universe is governed by such laws, it can be imagined, as Boscovich and Laplace did,[7] that God sets the initial conditions when He creates the universe and leaves it to the laws to do the rest.[8] Except for an occasional miracle, He no longer participates in His creation, since the laws do the work. Laws could be thought of as God's surrogates. During the eighteenth century, forces originating in matter replaced God as the immediate causes of motion. But even though God doesn't appear in explanations in physics, we continue to speak of laws "governing." Those physicists who think there is no governor don't take this literally but understand this as a metaphor.

One may wonder, if laws are not God's volitions, what are they? and if governing is just a metaphor, how are laws enforced? John Foster argued that if there is no God, there are no laws and used this to argue for God's existence.[9] Nancy Cartwright agrees with Foster's conditional, but reversed the inference to argue against the existence of laws.[10] Her view is that if the theological assumptions that provide the underpinning of the idea of laws are no longer taken seriously, the concept should be abandoned.[11] Unlike Cartwright, I think that the idea of laws of nature is required to understand physics and other sciences and that it also plays a central role in understanding the metaphysics of space and time, fundamental ontology, counterfactuals, causation, chance, and many other concepts important in philosophy of science and metaphysics. Unlike Foster and Cartwright, I think the concept of laws of nature can survive without relying on theological assumptions.

The seventeenth-century search for fundamental laws brought with it accounts of fundamental ontology that satisfy the laws and a

---

[7] See Kiznjak (2015) for a discussion of Boscovich and Laplace on determinism.
[8] Laplace famously asserted that determinism is true and when asked by Napoleon what role the Creator has in his system replied that he has no need for that hypothesis.
[9] Foster (2004).   [10] Cartwright (2005).
[11] In Cartwright (1983), (1989), and (1999), she proposes and develops an account closer to the Aristotelian view on which the activities of diverse powers result in nature's regularities.

fundamental space-time arena that the ontology occupies. For one version of Newton's theories, the fundamental ontology consists of corpuscles or atoms and the space-time is absolute Euclidian space and absolute time. Other proposals were debated in Newton's time and after. The important point is that fundamental laws, ontology, and space-time were introduced together in proposals for fundamental theories of the world whose aim is to account for the motions of macroscopic objects.

The idea of chance also emerged during the sixteenth and seventeenth century. It too has precursors going back to ancient times. Greek and Roman atomists said that the motions of atoms are subject to occasional "swerves" attributed to chance. Later thinkers saw God's hands in the activity of chance in ways that arrange for stable long-term frequencies even though it is impossible for humans to accurately predict individual events.[12] The first rigorous applications of a mathematical concept of chance were to gambling games, actuarial tables, agriculture, finance, and other mundane matters that seem far removed from the celestial motions that were the paradigm domain for Newton's laws. For much of the seventeenth and eighteenth centuries the spheres of laws of nature and chance hardly overlapped, and chance played little role in physics. In fact, events that happen by chance were sometimes thought to be not covered by law at all. But by the twentieth century, chances are found throughout the biological and social sciences, and in physics laws and chance were joined in statistical mechanics and quantum mechanics. Laws and chance came to be understood as complementary rather than conflicting.

But how are laws and chance related? If given F, the chance of G is x, then the chance of G given F seems to be a degree of lawful necessitation. What can that be? If the fundamental laws are deterministic, as they are thought to be according to classical mechanics,[13] the degree to which the complete state at a time lawfully necessitates future states is 1 or 0. What room does this leave for intermediate valued chances?

---

[12] The fact that the frequency of the number of male births to female births per year is constant even though it is impossible to predict the gender of the next birth was taken to be evidence for God's existence.

[13] There are complications about exactly how to characterize "determinism" and then whether classical mechanics is exactly deterministic. See Earman (1986) for discussion.

These questions are discussed in Chapter 6 and more thoroughly in the 2020 sequel to this book, *The Mentaculus Vision*.[14]

Ian Hacking gives the birth year of the concept of chance as 1654—the year of a famous correspondence between Pascal and Fermat that concerned how to settle the winnings in a game of dice interrupted by the king's gendarmerie.[15] Solving this problem involved a concept that, as Ian Hacking puts it, is "Janus faced."[16] One face looks to the world and the other to the mind. Chance says something about the world—the outcome of a toss of dice—and also something about the degrees of belief and corresponding rational betting odds one ought to have for the outcome. Some accounts of this concept characterize it in terms of one or the other aspect and then try to explain, or explain away, the other aspect. But as Hacking emphasizes, the two faces are inseparable. Chance is an objective feature of the world that guides rational subjective belief and action.

The primary philosophical question about chance is "What feature of the world can be like that?" A related question is "How is chance related to laws of nature, and in particular, what room is there for chance if the fundamental laws are deterministic, since such laws perfectly predict future events?" The account of laws and chance I will offer in this book provides answers to these questions by unifying the metaphysics of laws and chance.

While it is the job of physics and the other sciences to find what laws and chances there are, it is the job of philosophy to find out what laws and chances are. This book examines the most prominent philosophical accounts of what they are and offers a new account. It goes on to show how the new account engages with theories in contemporary physics and with central issues in metaphysics. *The Mentaculus Vision* applies the new account to a framework for a physical theory and uses it to answer philosophical questions about the foundations of statistical mechanics, counterfactuals, causation, arrows of time, rational belief, and free will.

If not God, then what, as Stephen Hawking asks, is it that "breathes fire into the equations"? My attempt to answer Hawking's question first

---

[14] In preparation.   [15] According to Hacking (1984) this story is mythology.
[16] Hacking (1984) 42.

involves surveying the most prominent contemporary accounts of laws and chance. I tentatively come down on the side of David Lewis's Humean Best Systems Account (BSA). I will then propose a novel account that builds on the BSA. My account, the Package Deal Account (PDA), differs from the BSA in that, while the BSA is built on a metaphysically given ontology of fundamental properties/quantities and space-time that is compatible with Humean metaphysics, the PDA has no such metaphysical presuppositions. It incorporates laws, chances, fundamental ontology and properties, space-time, and the special sciences into a package of mutual interdependence that is not metaphysically presupposed.

Like Lewis's BSA, the PDA claims that a proposition expresses a law not in virtue of governing, but by its role in a scientifically optimal systematization. However, it is not committed to Lewis's version of Humean metaphysics and its rejection of necessary connections. Because of this it avoids some of the problems that confront Humeanism, better fits some contemporary theories in fundamental physics that posit unusual ontologies and space-time structures, can include elements of other accounts, and replaces metaphysical assumptions with scientific theory. It advances the project of "naturalizing" metaphysics by fitting in with a broadly Quinean epistemology of science in a way that overcomes certain skeptical challenges to the existence and knowledge of laws. At the same time, it is compatible with scientific realism and even with a kind of metaphysical realism. I will argue that the PDA better captures how laws, chances, and fundamental ontology are understood in physics than any of the other accounts on offer.

It is controversial whether, in addition to the laws of physics, there are genuine laws of special sciences like, thermodynamics, geology, biology, psychology, economics, and so on and, if there are, how they are related to fundamental laws of physics. But it is clear that there are special science regularities, including probabilistic regularities and ceteris paribus regularities that are involved in explanation and confirmation, whether or not they are called "laws." The PDA, by including the special sciences in its account of laws provides an account of how special science regularities and probabilities are related to fundamental physical laws and probabilities. However, most of the ensuing discussion of the metaphysics of laws concerns the fundamental laws of physics.

2
Non-Humean Accounts of the Metaphysics of Laws

The concept of fundamental law of nature that emerged during the seventeenth and eighteenth centuries contained two main ingredients. One is that laws are expressed by simple mathematical principles that unify and systematize fundamental events. The second is that laws describe how God governs these events so that they conform to His will. In the seventeenth century, the systematizing involved in laws was thought to reflect Divine nature, and the governing was thought to be based on God's power and activity.[1] During the ensuing centuries, the theological underpinnings of the concept of law of nature fell away, and the notion of law spread from physics to the special sciences like meteorology, geology, chemistry, biology, economics, etc. But exactly how this happened is not well understood. The historian Frans van Lunteren has said that "the history of the concept of laws of nature has yet to be written."[2]

Contemporary approaches to the metaphysics of laws and chances dispense with theology while emphasizing one or the other of these two ingredients. The approaches are divided into two camps, called "Humean" and "non-Humean." Contemporary Humeans emphasize the systematizing and unifying aspect while rejecting governing, and contemporary non-Humeans emphasize governing and necessitating without relying on theology. The two approaches differ in that non-Humean accounts endorse fundamental necessary connections either between governing laws and regularities or between properties connected by law. Humean approaches follow Hume, the great denier of fundamental

---

[1] See Ott (2019) and (2022) for accounts of who shows how both systematizing and governing are central to Berkeley's conception of law of nature.
[2] Van Lunteren (2016). However a recent book by Walter Ott (2022) contains an extensive discussion of this history.

necessary connections. They maintain that the apparent necessity of laws derives, at least partly, from us.[3] This chapter addresses non-Humean accounts, and the subsequent chapter addresses Humean accounts and focuses on David Lewis's BSA.

There are two main kinds of non-Humean accounts: (i) governing accounts and (ii) powers accounts.[4] Governing accounts hold that fundamental properties are categorical meaning that they are individuated independently of laws and causation. They construe laws and chances as involved in elements of reality over and above the distribution of fundamental categorical properties[5] that in some way govern, constrain, or guide it. Fundamental necessity is involved either in characterizing laws or characterizing governing or both. Powers accounts hold that some fundamental properties are not categorical but are dispositions or powers whose activities result in lawful necessary connections. The instantiation of a disposition or power at one point produces or has a tendency to produce an instantiation of property at another point. When it does, these instantiations are necessarily connected. Laws are the regularities that result.

Non-Humean accounts of chance describe it as a measure of a degree of propensity or a power. Either laws or powers determine the chances. Chances so understood do not govern but rather somehow guide the evolution of events. If it is a law that if an F occurs, the chance of E occurring is x, then, while there is no guarantee that E occurs, x measures the degree to which F tends to lead to E.

---

[3] Humean accounts need not reject necessary connections among mathematical facts or derived from conceptual connections.
[4] There is also a non-Humean account due to Marc Lange that differs from governing and powers accounts in certain respects. It takes counterfactuals as fundamental and characterizes laws in terms of them. The rough idea is that laws are propositions that are true and remain true under various counterfactuals suppositions. The account is very ingenious and has certain advantages over other non-Humean accounts. But it is also subject to some of the same problems I raise for governing and powers accounts. Further, by making counterfactuals fundamental and laws derivative on them it raises the question of what makes counterfactuals true. His account seems to put the counterfactual cart in front of the nomological horse. See Lange (2009) for an exposition and Loewer (2011) for criticism of this approach.
[5] Lewis calls the distribution of fundamental categorical properties and relations throughout all of space-time "the Humean mosaic."

Governing Accounts

Contemporary governing accounts of laws replace the Deity with something else that does the governing. David Armstrong proposed necessitation relations between universals for the job.[6] The universals involved in laws are entities that are or determine properties which are not themselves individuated in terms of laws or necessary connections relating instances instantiated in different space-time regions. Universals or properties that are not individuated in terms of nomological or causal connections are called "categorical."[7] According to Armstrong, the generalization "Every F is followed by a G" expresses a law iff F and G are categorical universals that are related by "contingent necessitation" N(F,G). Given the occurrence of an F, N(F,G) produces the occurrence of a G. Armstong says that N(F,G) is contingent because it is metaphysically possible for F and G to be instantiated even though they are not related by N. However, according to Armstrong it is not metaphysically possible for N(F,G) to obtain but $(x)(F \to G)$ to be false.

Since N(F,G) guarantees that Fs are followed by Gs, it is supposed to explain why Fs are followed by Gs. Bas van Fraassen calls the problem of demonstrating that N(F,G) implies $(x)(Fx \to GX)$ "the inference problem."[8] Armstrong does not see how to solve it but simply accepts that it is the case, as he says, "with natural piety." Thus, N(F,G) replaces God governing the world.[9]

Armstrong's account succeeds in distinguishing lawful from accidental regularities, but it faces two big problems. The first is that the laws in contemporary physics like Hamilton's equations, Schrödinger's equation, and the field equations of general relativity do not relate Armstrong-type universals and do not take the form of material

---

[6] Armstrong's account is in *What Is a Law of Nature*. Around the same time similar proposals were made by Fred Dretske (1977) and Michael Tooley (1977).

[7] Universals are entities that determine properties. Lewis understands properties as sets of possible individuals.

[8] Van Fraassen (1989).

[9] One suggestion for solving the inference problem is simply to make it a feature of N(F,G) that when it obtains $(x)(Fx \to Gx)$ is true; see Schaffer (2016). But this doesn't really solve the problem since it just raises the question of what kind of relation can be like this and why believe there are any instances of it. See Ioannidis et al. (2021) for further criticism of Schaffer's suggestion.

conditionals. Instead, they are expressed by differential equations that specify functions among various quantities. Dynamical laws specify how the entire state of an isolated system (or the entire universe) evolves or is constrained. Dynamical laws are what Tim Maudlin calls "FLOTES"—fundamental laws of temporal evolution—that specify the evolution of the fundamental state of a system.[10] These states are not Armstrong's universals. A fundamental state specifies which fundamental quantities are exemplified at each spatial point at a time. Armstrong could take this into account by stipulating that the relation of contingent necessitation can hold between entire states of the universe or states within a light cone, but this is far different from his original proposal.

A more philosophical problem with Armstrong's account was alluded to before. It is that he provides no account of what N(F,G) is and how it is that it makes it the case that Fs are followed by Gs. One could insist that it is simply analytic that if N(F,G) then Fs are followed by Gs. But this undermines the ability of N(F,G) to *explain* why Fs are followed by Gs, in the same way that calling a drug a soporific fails to explain why it causes sleep. So, what is the relation? It is not the case that N(F,G) *causes* Fs to be followed by Gs. N(F,G) is not the sort of entity that is involved in causation; since causation involves subsumption under laws this would require laws involving N(F,G) and so would just postpone the problem anyway. For N(F,G) to guarantee and explain why Fs are followed by Gs it would have to involve some kind of sui generis metaphysical relation. It is this kind of fundamental necessary connection that Humeans reject. But where Descartes, Newton and their contemporaries might have thought they understood how God could bring about worldly regularities, we have to admit that we have no idea how N(F,G) works to make Gs follow Fs, or explains why they do.

The burden is on Armstrong to show that N(F,G) possesses the features of laws. Among these features are that laws form part of a system, support counterfactuals, and underlie explanations. Just declaring that N(F,G) relates F and G by a relation called "nomological necessity" doesn't show that it does these things. N(F,G) has to earn its status as a law, not merely assume it. As Lewis quipped, Armstrong doesn't obtain

[10] Maudlin (2007).

strong biceps merely by being named "Armstrong." N doesn't make for laws merely by being named "contingent necessitation."

A different governing account of laws that is more in tune with contemporary physics has been proposed by Tim Maudlin.[11] According to Maudlin, the fundamental laws of physics are themselves metaphysically fundamental elements of ontology. They are entities, in a broad sense, although they don't fall into any of the usual ontological categories, such as properties, relations, individuals, fields, space-time, abstract mathematical structures, etc. Maudlin thinks we should recognize laws as belonging to a *new* category of ontology. A law apparently occupies all of space-time and produces or *governs* and thereby *explains* the evolution of events in space-time. They do this by operating on the state of the universe at a time (or on a Cauchy surface) and producing from it states at subsequent times. This producing is not ordinary causation since causation is between local events. Accounting for why the existence of a law makes instances of a regularity obtain is the version of the inference problem for Maudlin's account. Exactly how it works is not further explained but posited as the best account of laws in physics.

Maudlin's account has the interesting feature of presupposing a preferred temporal direction in which laws operate. He thinks this is an advantage of his account since he has other reasons to hold that, as he puts it, "time passes" in the direction of past to future. On his account, time is unique among the four dimensions of space-time in being imbued with an intrinsic direction. According to Maudlin, this direction not only determines the direction of the operation of laws but also underlies other manifestations of time's arrow—for example, that we can influence the future but not the past, that there are records of past events but not future events, and that we experience time as flowing from past to future. Thus, in contrast to Armstrong's account, Maudlin's involves an intimate relation between laws and time and, more generally, with spatio-temporal structure, since FLOTES on his account reflect this structure by exhibiting space-time symmetries.

There is a problem adapting Maudlin's account to relativity, since on usual understandings relativity rejects the existence of the state of a

[11] Maudlin (2007).

system at a time. This means that Maudlin's account requires accepting a particular foliation of space-time as ontological fundamental. The laws then operate on the state at a time determined by a foliation. He is willing to accept this since he holds that a preferred reference frame is also required by Bohmian quantum mechanics, which he also accepts. General relativity presents the further challenge of apparently permitting time-like loops in which case there is no state at a time at which the laws can operate. Maudlin's response to this is to declare that space-time structures that admit time-like loops are metaphysically impossible.

Maudlin's account is also incompatible with cosmological models like Sean Carroll's parabolic universe, in which entropy increases without limit from a low "middle" point in both temporal directions, if the direction of time is understood as specifying the direction of increase of entropy.[12] Events proceed in both directions from the low entropy middle as though it is in the past and there are two futures proceeding from it. But Maudlin's arrow only goes in one direction, so which way does it go?

Wüthrich and Lam argue that neither Maudlin's, nor any other account of laws that presupposes space-time is suitable for quantum gravity theories which treat time and space as non-fundamental and emerging from something else.[13] For example, loop quantum gravity takes time and space as emergent from spin networks. These networks satisfy certain lawful conditions that don't involve space-time but can be configured by laws so as to give rise to space-time. These laws cannot be Maudlin's FLOTES.[14] The account of laws I specify later will be able to accommodate theories with ontologies that don't include space-time as fundamental.

Chen and Goldstein have articulated an anti-Humean account of laws which is similar to Maudlin's in certain respects but doesn't involve any metaphysics of time and thus avoids some of the problems Maudlin's

---

[12] Carroll (2010).
[13] See Wüthrich and Lam (2023) for discussion of similar concerns about Maudlin's account.
[14] Although I am being critical of Maudlin's account of laws I think it is the clearest and most forceful version as an anti-Humean account that is compatible with the way contemporary physicists think about laws. Reflecting on Maudlin's account and his criticisms of Humean accounts is what prompted me to write this book.

account faces.[15] According to their view, fundamental laws are or are components of global constraints on the universe that determine what is nomologically possible.

Chen and Goldstein don't initially propose any conditions on the propositions that can count as laws. As a result, their constraint view is more flexible than Maudlin's and Armstrong's accounts, since it allows for the existence of laws, like the field equations of general relativity, that don't relate universals or even the state of a universe at a time to subsequent states, as well as laws like the Past Hypothesis (PH)[16] that is nondynamical. But just because their account provides no conditions except that laws constrain the universe, it also leaves open the metaphysical possibility that there is only a single law—for example, that all crows are black. In a world with just this constraint, all other regularities are accidental. They make their account more plausible by suggesting that the law yielding constraints determine a system of the world that optimally balances informativeness and simplicity. In this way, like Descartes's account, their account combines both the governing and systematizing aspects of laws. But where Descartes had an explanation of why these two aspects are connected, Chen and Goldstein just glue them together.

Since constraints are expressed by propositions, one might think that their view is that laws are propositions. But they are not just any old propositions, since, unlike most propositions, they constrain the universe. There is something in reality that makes these propositions constraints. Their view raises questions similar to questions that confront the other governing views. Exactly what makes a proposition a constraint? And how does it do its constraining? For them, constraints and constraining must be understood as primitive. We shouldn't think that because we understand what it is for chains to constrain the motions of a wild dog, we understand how propositions constrain the universe in a way that enables them to do the work of laws. "Constraint" is a metaphor as much as governing.

---

[15] Chen and Goldstein (2023). Their view allows for the possibility that temporal arrows are reducible to the constraining laws and the contents of space-time.

[16] The Past Hypothesis describes the macroscopic state of the very early universe as one of very low entropy. It is needed in an adequate account of statistical mechanics as applied to the whole universe; see Albert (2000).

Chen and Goldstein's account is an improvement in some ways over Maudlin's, which is an improvement over Armstrong's, as accounts of fundamental laws of physics. All involve identifying laws with something over and above the events that laws are supposed to govern or constrain and a relation of governing or constraining that is supposed to explain these events, support counterfactuals, and so on. These items take the place of God and His governance assumed in the eighteenth-century view of laws.

Powers Accounts

The second type of non-Humean account is "powers accounts." These are, in certain respects, throwbacks to the Aristotelian view of science that the seventeenth-century account based on mathematical laws replaced.[17] On powers accounts, it is the nature of some properties to possess the power of making its instances produce appropriately related instances of properties which themselves have the capacity to be produced. Contemporary advocates of powers accounts, like Alexander Bird,[18] completely reject governing laws in favor of powers as the source of motion. On their view, laws are not fundamental elements of reality but are regularities that result from the behavior of powers. Powers do the producing, and laws are the regularities that result.[19] This proposal may be a way of making sense of the governing metaphor.[20] Although on the powers view lawful regularities are metaphysically necessary, the contingency of laws is preserved since the fact that a power F is instantiated is contingent.

---

[17] Some proponents of governing laws kept a diminished role for something like powers. For example, Newton apparently thought that matter possesses the power or capacity to follow the laws of motion and gravitation; see Psillos (2018).
[18] Bird (2005) is one of the primary proponents of a powers account of laws.
[19] One way to develop this idea is that there are a limited number of fundamental physical entities and properties upon whose powers the powers and lawful regularities of non-fundamental reality depends. Nancy Cartwright is also a proponent of fundamental powers, but she rejects that they are limited to fundamental properties and holds that higher-level properties may have their own powers and capacities that are not derivable from those of fundamental powers. She also holds that the lawful regularities powers produce may only hold ceteris paribus.
[20] See Psillos (2006) and (2014) for discussions of powers and laws.

One of the problems with powers is that they, like Armstrong's universals, are out of tune with contemporary physics. FLOTES like the law expressed by Schrödinger's equation relate the entire state of an isolated system at one time with its state at other times, while the powers account connects instantiations of powers with other instantiations of powers. One could derive state-relating laws from powers, but that would require composition principles in addition to powers and these principles would need to be laws. Also, it is difficult to see how this can work for quantum mechanics, because the quantum state of a system is not composed of the quantum states of parts due to entanglement. I suppose one can stretch a powers account to accommodate this by saying that the entire state at a time (which might be the state of the universe) has the power to produce subsequent states.[21] But this is a far cry from the original proposal.

There are a few further points to be made about powers accounts. First, like Maudlin's account, they assume a metaphysically fundamental direction of time. F's power is for its instances at one time to bring about a G at subsequent times. We can remove the presupposition of a primitive direction time by dropping talk of powers and just propose the view that the instantiation of a property at one time (or location) metaphysically necessitates the instantiation of properties at distinct times and locations. This is like Shoemaker's proposal that some properties are individuated in terms of their necessary connections with other properties and laws are the regularities that result.[22] On this account, fundamental properties need not be understood as powers whose instantiations "produce" subsequent instantiations of properties. But this removes the "oomph" that anti-Humeans think is required for lawful explanation and, as we will see later, it may not even be incompatible with a certain kind of "Humean" account of laws.

---

[21] The parts of a quantum system may not have quantum states of their own. A part can be described by a density matrix, but because parts may be entangled with other parts the quantum state of the system is not given by the density matrices of its parts. Walter Ott (2022) proposes a power-like view on which the entire state of the world possesses the power to produce subsequent states. It is difficult to see how this differs from governing accounts.

[22] Shoemaker's (1980) proposal is that some properties are individuated in terms of lawful or causal connections. This seems to presuppose a prior understanding of law or causation.

Second, if talk of powers is taken seriously, then the power of a property instance at one time to *produce* a property instance at subsequent times seems as mysterious as governing. Instead of a law making a G follow an F by necessitating the regularity that Gs follow Fs it is now in the power of F to make a G follow it and thus make an instance of F necessarily connected to an instance of G.

Non-Humean accounts possess an epistemic feature that may seem troubling. It is that there can be two possible worlds that seem exactly alike and yet differ radically with respect to their laws. For example, on a governing account there can be two worlds in which particles move on exactly the same space-time trajectories except that in the first the particle motions are governed by Newtonian laws, while in the second the trajectories are governed by different laws or no laws. On a powers account there can also be a world in which the particles move on the same trajectories but instantiate different powers. This is a consequence of the fact that on non-Humean accounts the laws and powers don't supervene on the totality of categorical property instantiations.

In these worlds counterfactuals differ, but we only have epistemic access to the actual facts. This raises an epistemological worry. The apparent epistemological problem is that if all we can directly see or test are the particle trajectories, then non-Humean accounts lead to skepticism. According to widely accepted interpretations, this was Hume's worry about fundamental causation. The alleged claim is that since all we ever experience are correlations among events there is no reason to believe that there is causation over and above certain correlations.

Proponents of non-Humean accounts of laws deny that their view leads to skepticism. They claim that the failures of laws to supervene on what they govern is an advantage of their accounts since they claim that we have intuitions that different laws or different powers can produce the same motions.

Further, non-Humeans claim that there is no epistemological problem since, although we cannot see powers or governing laws, we can make rational inferences about them because they provide the best way of explaining what we do see. If we lived in a world in which the particle trajectories satisfy Newtonian laws, but the actual governing laws are not Newtonian, it still may be that the best explanation of the trajectories is that they are governed by Newtonian laws and that these are governing or power laws. In this case an inductive inference may lead to a mistaken conclusion, but one it would still be rational to believe.

In fact, defenders of non-Humean accounts turn the tables and claim that the fact that Humean laws supervene on actual facts makes it the case that they are not capable of supporting explanations at all. If so, inference to the best explanation cannot be used to infer laws and thus results in skepticism about them. If that is correct, then it is the Humean who has the epistemological problem because much of our knowledge is based on inference to the best explanation. Of course, this conclusion assumes that non-Humean laws really can explain and Humean laws cannot, a point to which we will return.

3
The BSA Humean Account of the Metaphysics of Laws

David Hume is known for inspiring the regularity account of causation. According to this, a type of event C causes a type of event E when C precedes E and there is a correlation between occurrences of events of these types that satisfies some further conditions, and these further conditions don't involve fundamental necessary connections.[1] Hume himself doesn't say very much about the metaphysics of fundamental laws of physics even though it is certain that he was familiar with Newtonian mechanics and its laws and believed that causation involves lawful regularities. Despite the lack of discussion by Hume, the view that laws are correlations between types of events that satisfy further Humean conditions has come to be known as "Humean."

Hume is known as the great denier of necessary connections. For an account of laws to count as Humean, these further conditions cannot include any modal concepts like metaphysical or nomological necessity, counterfactuals, causation, or objective probability that are understood as referring to fundamental modal features of reality. If they are involved in an account of laws, they must themselves be explicable in non-modal terms. The problem for Humean accounts is specifying the additional conditions without adverting to fundamental modality that make a regularity lawful and showing that laws so characterized can perform the work of laws in science. This is a challenge since laws support counterfactuals, and counterfactuals seem to involve some kind of necessity or

---

[1] It may be, as Galen Strawson (1989) suggests, that Hume never rejects the existence of necessary connections and that he held that causation involved the obtaining of a necessary connection between C events and E events but that our knowledge is limited to the correlation. I won't venture an opinion about this since my interest is not in what Hume really thought but the views about laws he inspired.

modality. So, the problem for the contemporary Humean is showing how the kind of necessity involved in laws and counterfactuals can be built from ingredients of a Humean ontology rather than assuming that it is a fundamental feature of reality.

Necessity and possibility seem to be involved in laws since laws support counterfactuals, and counterfactuals seem to concern other possible worlds. For example, "if the match which has not been scratched were scratched it would have lit" seems to say something about how the match behaves in worlds that are not actual. Accidental generalizations lack the ability to support counterfactuals. As Reichenbach pointed out, "There is nowhere in the universe a densely packed one meter in diameter sphere of uranium" expresses a law, while a similar generalization with "gold" substituted for "uranium" is an accidental generalization. The first supports the counterfactual "if one were to try to produce a densely packed one meter in diameter sphere of uranium one would not succeed," while the second generalization involving gold does not support similar counterfactuals.

An early Humean proposal was that for a true generalization to be lawful its predicates, or the properties they denote, must satisfy a condition like being purely qualitative or denoting natural kinds. But this proposal is not satisfactory since it is easy to come up with true generalizations that connect predicates/properties satisfying these conditions that are not laws. Another proposal is that a law is a true lawlike generalization, and a generalization is law-like iff it is confirmable by its instances.[2] This is more promising since it has been argued that if a generalization is confirmable, it supports counterfactuals. This raises the question of what makes a generalization confirmable by its instances. It is generally held that being a candidate for being a law is what makes a generalization confirmable by instances. This suggestion seems to be putting the epistemological cart in front of the metaphysical horse. It also makes the notion of fundamental law an epistemological concept.[3] This the way some Humeans go but there is a better Humean way.

---

[2] This is Nelson Goodman's proposal in Goodman (1950).
[3] Goodman (1950) characterizes the confirmability of a generalization in terms of the projectability of the predicates that compose it, and the projectability of a predicate at a time in terms of actual successful projections. This results in the concept of law not only being epistemological but also relativized to time and histories.

A better proposal that avoids these difficulties was made by Mill and Ramsey and more recently developed by David Lewis. Lewis calls it "the Best Systems Account of laws" (BSA).

The BSA specifies which regularities qualify as fundamental laws in terms of how lawful regularities supervene on the non-nomological truths. Lewis didn't write a single paper devoted to his account of laws but describes it in several places. One is:

> Take all deductive systems whose theorems are true. Some are simpler and better systematized than others. Some are stronger and more informative than others. These virtues compete: an uninformative system can be very simple; an unsystematized compendium of miscellaneous information can be very informative. The best system is the one that strikes as good a balance as truth will allow between simplicity and strength. How good a balance that is will depend on how kind nature is. A regularity is a law iff it is a theorem of the best system.[4]

The guiding idea of the Best System Account is that what makes a proposition lawful is its participation in a scientifically optimal systematization of the distribution of fundamental properties throughout the entirety of space-time. According to Lewis a proposition Q is a law only if Q is a generalization entailed by a system that is the, or one among a number of, scientifically optimal systematization of non-nomological fundamental facts.[5] A fundamental fact is non-nomological iff it doesn't metaphysically entail any nomological fact. Since laws systematize non-lawful facts, they supervene on them. In other words, "It is a law that Q" does not state a fact over and above those facts but systematizes and thus unifies them.

Lewis identifies a law-determining best systemization as aiming to achieve an optimal balance between informativeness and simplicity. He thinks of informativeness in terms of possibilities excluded and

---

[4] Lewis (1994a) 478.
[5] Adding the condition that a lawful proposition is a generalization doesn't seem to add anything since any proposition can be formulated as a generalization and some sentences that are not generalizations like "The entropy of the early universe was very small" (aka "the Past Hypothesis") plausibly express laws.

simplicity in terms of syntactic and mathematical simplicity. He doesn't say much more about these criteria or how to balance them but seems to suggest they are determined by scientific practice.

Lewis immediately recognizes that his account requires a restriction on the language in which the candidates for best system are formulated. Without such a restriction the sentence VxSx, where S is a predicate true of all and only actually existing objects, counts as maximally informative since it excludes all possible worlds except for the actual world. It is also very simple and so would quality as an optimal systematization. Since VxSx entails every truth, the disastrous consequence is that every truth is lawful. To avoid this consequence Lewis proposes a restriction on the language in which candidates for optimal systems are expressed based on a central component of his metaphysics. According to it there are certain elite properties he calls "perfectly natural" which are the fundamental properties instantiated at possible worlds, and whose distribution in a space-time forms the supervenience base for all truths at that world.[6] These properties are instantiated at points (or small regions) and are categorical, i.e. the instantiation of one such property at a region does not metaphysically exclude or necessitate the instantiation of any other perfectly natural property at a distinct point (or non-overlapping region).

A perfectly natural property may be a magnitude in which case the value at one point excludes different values at the same point, but these are the only necessary connections among instantiations of perfectly natural properties. Lewis claims that one of the jobs of fundamental physics is to find perfectly natural properties/magnitudes and gives as possible examples charge and mass. But it is not a condition on being perfectly natural that perfectly natural properties/magnitudes are those that physics aims to find; rather, Lewis seems to believe that it is a condition on physics that it aims to find perfectly natural properties.

---

[6] Lewis can be understood as a substantivalist about the space-time arena that is composed of points satisfying geometrical relations. While he usually considers the arena to be four-dimensional and Euclidean there may be possible worlds whose arenas have different dimensions and different geometries. He can also be understood as a relationist about space-time. On this view the fundamental entities are individuals of some kind and there is a perfectly natural relation among them that gives rise to a space-time structure.

Lewis calls the distribution of the world's perfectly natural properties its "Humean mosaic" (HM).[7] We can think of the HM as a kind of field or overlapping fields of points whose values are perfectly natural properties/magnitudes. At each point natural property magnitudes are instantiated. In Lewis's Humean metaphysics all truths, including those involving laws, probabilities, counterfactuals, causation, explanation, everyday objects, etc. supervene on the world's HM.[8]

Lewis considers any generalization entailed by a world's best systematization to be a law at that world. But scientists would not call every such an entailment a law. For example, the disjunction of two laws would not be counted as a law. To earn the title "law" a sentence must play a role in scientific explanation and prediction. It is better for Lewis to say that every entailment of the world's best system is "lawful" and reserve the title "law" for those that play an appropriate role in explanation and other scientific practice.

Lewis proposes that the languages whose true propositions are systematized by a world's best system have only perfectly natural predicates in their extra-mathematical vocabularies.[9] Restricting the language in which candidates for best system are formulated avoids the disaster mentioned above, since Sx is not a perfectly natural predicate and its definition in terms of perfectly natural predicates is enormously complicated. But it does this at the cost of building the Best System Account on the back of the metaphysical posit of an elite class of fundamental categorical properties that don't derive their elite status in virtue of appearing in laws but, rather, certain regularities derive their status of being lawful in virtue of systematizing the distribution of instantiations of perfectly natural properties. I will return later to consider just how costly this posit is.[10]

---

[7] If the only perfectly natural relations that are instantiated at a world are geometrical relations, then Lewis says the world satisfies a condition called "Humean Supervenience." He speculates that the actual world may satisfy this condition but worries that it may be violated by quantum mechanics. For a discussion see Loewer (1996).
[8] This is a very bold metaphysical conjecture. It entails, as an instance, that a true history of the Gallic Wars supervenes on the actual Humean mosaic. A great deal of Lewis's work is devoted to making this plausible.
[9] We will see later that he relaxes this requirement to allow probability relations to appear in extra-mathematical vocabulary and we will examine a proposal for relaxing it further.
[10] One can think of restrictions on languages for candidate best systems other than Lewis's that are not are metaphysically committed.

Another worry about the account is that since laws are characterized in terms of informativeness and simplicity and these seem dependent on human cognitive abilities, interests, and practices it seems to make the concept of law also dependent on human interests and practices. Lewis worries that this commits his account to what he calls "rat bag idealism." As the name suggests, he consider this to be a bad consequence of his account if there is no way to mitigate it. I will return to this later as well.

Having settled the issue of the language whose truths are to be systematized, the next issue is how candidate systems are to be evaluated. Lewis's account here is sketchy and inadequate. His measure of informativeness in terms of the size of the set of possible worlds excluded by a proposition is much too crude. Since there are infinitely many possible worlds, it restricts comparisons of informativeness to sentences that are related by necessary implication. Further, it counts logically equivalent sentences as equally informative even when one is much more complicated than the other. But scientists are interested in not just a quantity of information measured in terms of possibilities excluded but also in how the information is presented and organized in devising law-determining systems.

Several authors have argued that law-determining systematizations should aim to organize information that is important to scientists in ways that are usable to them especially for prediction, explanation, and support of counterfactuals and causal connections.[11] A suggestion along these lines was made by Ned Hall.[12] He observed that a systematization that supports a distinction between states of a system (or the universe) at a time and dynamical laws is much more useful for explanation and prediction than a system that doesn't make this distinction. The distinction between initial conditions and dynamical generalizations seems to be objective and independent of particular interests and abilities.[13] This is certainly correct and should be added as a condition that a lawful system should balance.

---

[11] See e.g. Dorst (2019); Loew and Jaag (2020).   [12] Hall (2015).
[13] James Woodward (2018) also stresses the importance of a system of laws distinguishing between initial conditions and laws.

A reason for requiring that lawful systems have a way of distinguishing states from lawful dynamical regularities is that this division plays a central role in characterizing counterfactuals. A good system should attempt to maximize not only information about what actually occurs but also information about what would occur under alternative circumstances. This is a point we will return to later when discussing counterfactuals and laws.

Another criterion is that an optimal systematization of a Humean mosaic is also approximately an optimal systematization for typical, sufficiently large portions of the mosaic. This enables application of laws to subsystems. I propose that we add to Lewis's account the two additional criteria: (i) that it counts in favor of the informativeness of a system Σ that it includes dynamical laws but excludes initial conditions; and (ii) that it also best systematizes typical, sufficiently large subsystems.

Lewis proposes the criterion "fit" for evaluating probabilistic theories. The fit of a theory to a mosaic is given by the probability of the mosaic given the theory. By entailing probabilities of propositions, a theory provides information about the mosaic. Exactly how it does this, as well as other issues and problems with Lewis's account are discussed in Chapter 9.

Lewis's proposal regarding simplicity is also inadequate. He suggests the length of sentences expressed in terms of perfectly natural predicates as a measure of simplicity. But this doesn't have anything to do with the way physicists evaluate simplicity when considering alternative theories. Physicists prize symmetries, small numbers of fundamental quantities, small numbers of parameters and constants, their relationships, and so on.

In addition to what might fall under the headings "informativeness" and "simplicity," there are other criteria that physicists require fundamental theories to satisfy. Following Einstein and Jeans it has become usual to make a distinction between principle and constructive theory components of fundamental theories and laws.[14] There is a corresponding

---

[14] See Lange (2014) for a discussion. Distinguishing among these components of a fundamental system allows for an account of counter legal counterfactuals. This will be discussed in Loewer (forthcoming b).

distinction between kinematic and dynamical laws, and a requirement that they harmonize. Requiring that dynamical laws enforce certain space-time symmetries accounts for the special role of simple conservation laws like the conservation of momentum and energy.[15] So I suggest adding informativeness to Lewis's criteria of simplicity, and fit the further criteria of making a distinction between states and dynamical laws, laws applying to subsystems, and making a distinction between principle and constructive components. An optimal system optimally balances these and whatever other criteria are used by physicists when evaluating proposed fundamental theories.

Much more needs to be said about the criteria for evaluating systems. But even if Lewis and I have not satisfactorily characterized these criteria, we can see that he is gesturing at criteria that are employed within physics to evaluate theories, and especially proposals for fundamental theories. I assume that there are such criteria guiding theory choice when physicists evaluate proposals for fundamental theories, even if these criteria are evolving and not yet settled. Their evolution is guided by the aim of achieving an account of the universe that enables the systematization of the world so as to yield a usable system for making accurate predictions and, most importantly, unifying explanations. We can see why satisfying these criteria is valuable to physics and so why the discovery of laws is valuable. For now, I will take Lewis's "informativeness" and "simplicity" as place holders for whatever criteria physicists try to balance in evaluating proposals for fundamental theories.

By requiring that a best system of the world satisfies these criteria the BSA helps explain why mathematics is so important to the formulation of laws of physics. Mathematical propositions are able to express a great deal of information simply. Furthermore, the BSA enables discovery of the consequences of laws when applied to initial conditions and subsystems. Descartes thought that fundamental laws are mathematical because God is a mathematician. But the BSA explains why laws are mathematical independently of that hypothesis: systems that optimize satisfying the criteria that determine lawfulness are mathematical.

---

[15] This is demonstrated by Noether's theorems.

A vivid and useful, if fanciful, description of the BSA is provided by David Albert who, somewhat ironically, imagines an interview with God:

> You get to have an audience with God. And God promises to tell you whatever you'd like to know. And you ask Him to tell you about the world. And He begins to recite the facts: such-and-such a property (the presence of a particle, say, or some particular value of some particular field) is instantiated at such-and-such a spatial location at such-and-such a time, and such-and-such another property is instantiated at such-and-such another spatial location at such-and-such another time, and so on. And it begins to look as if all this is likely to drag on for a while. And you explain to God that you're actually a bit pressed for time, that this is not all you have to do today, that you are not going to be in a position to hear out the whole story. And you ask if maybe there's something meaty and pithy and helpful and informative and short that He might be able to tell you about the world which (you understand) would not amount to everything, or nearly everything, but would nonetheless still somehow amount to a lot. Something that will serve you well, or reasonably well, or as well as possible, in making your way about in the world. And what it is to be a law, and all it is to be a law, on this picture of Hume's and Lewis's and Loewer's, is to be an element of the best possible response to precisely this request—to be a member (that is) of that set of true propositions about the world which, alone among all of the sets of true propositions about the world that can be put together, best combines simplicity and informativeness.[16]

I would only disagree with Albert's story by modifying it so that you don't ask God to provide information just for you in your particular situation, or only for the purpose of your making your way about. That is too limited and not the aim of science.

Albert emphasizes the pragmatic aspects of the BSA. In my view some recent discussions of Humean accounts of laws have overemphasized

---

[16] Albert (2000) 23.

these pragmatic elements.[17] On such pragmatic accounts what makes a system law determining is its usefulness in enabling predictions of phenomena that interest scientists. An overly pragmatic understanding of the BSA disqualifies the field equations of general relativity and the laws of quantum field theory and much else considered to be part of fundamental physics from counting as laws. These laws contain a lot of information about fundamental ontology and provide understanding of how it is structured but a great deal of it is not relevant to what we ordinarily mean by "getting about in the world," and is in a form that is useless for pretty much any practical purpose, at least without a great deal of mathematical massaging.

A better version of Albert's story has you asking God for a system of information which a sufficiently sophisticated and educated human, wherever she is in the universe, can use not just to help "make her way around in the world" but more importantly to explain what she finds there in terms of fundamental ontology and fundamental laws. She should ask for a system that unifies, to the greatest extent possible, fundamental facts and facts of the special sciences. In light of this, the system that God gives her may be useless for enabling her to predict matters that concern her like the weather, the stock market, preventing the flu, and so on. For such matters, regularities expressed in a macroscopic vocabulary that may hold only for the most part are much more useful. What she wants from God is an optimal systematization of the Humean mosaic that organizes and unifies facts so as to enable her to understand the world. This system will enable her to understand *why* the more useful regularities obtain. That is, she should ask God to provide her with something very much like the kind of system that Descartes thought describes the fundamental principles God follows to govern the world and what physicists later came to describe as a "theory of everything." Advocates of the BSA think that these principles do all the work science requires of fundamental laws but don't govern as non-Humeans claim.[18]

---

[17] Pragmatic understandings of the BSA have been proposed by Hicks (2018), Loew and Jaag (2020), and others.
[18] See Blanchard (2023) for a more extensive discussion of a similar account of the aims of a best system.

The BSA and non-Humean accounts have very different relations to time and existence. First, the BSA presupposes eternalism and a block universe, since the laws supervene on the entire Humean mosaic that includes events at all times—present, past, and future. This is the so-called "block universe."[19]

Lewis assumes a block universe with three spatial and one temporal dimension and a Euclidian geometry. But the BSA can be easily extended to space-times with non-Euclidian geometries in which the temporal and spatial dimensions are not completely separate, such as relativistic space-times, and to many dimensions, as in some versions of quantum mechanics, and even to theories that don't assume a fundamental space-time at all, like some recent proposals in quantum gravity theories.

Non-Humean accounts of laws do not presuppose eternalism or a block universe and are compatible with presentism and growing block[20] accounts of time and also compatible with eternalist accounts of time. On a governing account like Maudlin's (Maudlin is an eternalist), a law can be thought of as operating on a state of the universe and bringing about subsequent states to come into existence. Same for the powers account, on which the instantiation of a power produces subsequent instantiations. Although these accounts are compatible with eternalism, they do presuppose a fundamental temporal dimension with directionality.

The BSA provides an account of laws that dispenses with governing and powers. While it relates to the epistemology of fundamental physics by its reliance on the criteria that physicists use for selecting among candidates for laws, it is a metaphysical, not an epistemological account of laws. It is compatible with physicists being wrong about what laws there are.

---

[19] It is sometimes said that a block universe is static and that in it there is no change. But this is a confusion. Something is static if it doesn't change over some temporal interval. But the block universe is not in time, time is in it. Change occurs in the block when the block is different at different times.

[20] Presentism is the metaphysical view that only events that occur at a single time, now, exist while the growing bloc view holds that the present together with what is past exist. They are obviously incompatible with the BSA, since on it the laws supervene on what is past, present, and future. There is some more discussion of these metaphysical views about time and their connections with chance in Chapter 8.

4
Objections to the BSA

Is the BSA adequate? Non-Humeans are incredulous that mere descriptions, no matter how simple, informative, and unifying, can do the work of laws. They think that the laws add "oomph" to the universe. Before discussing these objections, I want to clarify a few points in favor of the BSA.

The first point in its favor is that, like non-Humean accounts, it counts some, but not all true generalizations as lawful. It thus avoids the objection to simple regularity accounts that they consider all true generalizations to be lawful. Second, it has the advantage that it specifies lawful true generalizations in a way connected with scientific practice, since satisfying the criteria for evaluating law determining systems is desirable in a scientific theory. If a physicist finds a generalization that she has reason to believe is true and entailed by the (or a) system she thinks is scientifically optimal, she would think that it is a law. Third, as I mentioned, although simplicity and informativeness are appealed to in evaluating the acceptability of a proposal for a system of laws, Lewis's account is not epistemological but metaphysical. The fact that a proposed candidate system compatible with the evidence is simple and informative is reason to think it is a law determining best system but the fact that it is a system that optimally balances simplicity and informativeness is what makes it law determining. On some epistemological accounts of law, it is the fact that a generalization is confirmable by its instances that makes it a law, but on the BSA, being a good candidate for a BSA law is what makes it confirmable by its instances.

Another point in favor of the BSA is that, according to it, what makes a proposition express a law is its role in an optimal systematization. I will argue later that laws specifying probabilities inform by recommending appropriate credences about mosaic facts.

Additionally, the BSA allows for laws with exceptions or a ceteris paribus qualifier, and vague laws. Markus Schrenk (2007) notes that some generalizations are considered to be laws, even though they are strictly false since they have certain exceptions. His—controversial— example is that the dynamical laws of physics are said to not apply within black holes. Such law statements are common in the special sciences. If indeed there are such laws, the BSA can account for them as being entailed by the best system.[1]

Eddy Chen has argued that some laws are vague.[2] His main example is the Past Hypothesis that specifies conditions of the early universe including its very low entropy but there are other examples, especially in the special sciences, like economics, that also are vague. Chen argues that the Past Hypothesis is a vague specification of an early universe macro state. This doesn't present a problem for the BSA since candidates for a best system may contain vague propositions. It is not as easy to see how anti-Humean accounts can do the same.

Lastly, an optimal system may include axioms that specify that certain symmetries are satisfied. The laws expressed by such axioms aren't dynamical but rather provide conditions that dynamical laws must satisfy. A best system also might include axioms that specify initial conditions and axioms that include macroscopic vocabulary. Allowing for this modifies Lewis's version of the BSA. Later, we will see that this is important in an account of the laws of statistical mechanics.

Since the BSA completely leaves out governing, necessary connections, and their ilk from the conditions that make a generalization lawful it must be admitted up front that the BSA should not be thought of as a *conceptual analysis* of the concept of law of nature as it has been handed down from the seventeenth century and is currently used unreflectingly by scientists and analyzed by some philosophers. It is better to think of it more along the lines of a proposal for refining and revising the old concept to produce a revised concept of law that is intended to adequately account for the roles that laws play in science. Such an account emphasizes the systematizing aspect of laws and is compatible

---

[1] See Schrenk (2007) and Chen (2022).   [2] Chen (forthcoming).

with Humean metaphysics. This doesn't make the BSA anti-realist about laws any more than a compatibilist account of free will is anti-realist about free will. According to the BSA there are laws, but they are not what non-Humeans think they are.

I will begin by considering objections to the BSA by first addressing some that don't depend on rejecting Lewis's Humean metaphysics. John Roberts claims that the BSA confuses lawfulness with order. He points out that scientists do not consider all order in the universe as lawful or following from laws alone.[3] Roberts's example is the simple regularity that the planets all travel on elliptical orbits in the same direction around the sun. Astronomers do not consider this to be a law. He claims this makes a problem for the BSA since he thinks it is committed to counting any simple regularity like this one as a law.

My reply is that this understanding of the BSA takes Lewis's identification of the criteria for evaluating candidates for scientifically best system as informativeness and simplicity too literally. Roberts applies it to individual regularities while Lewis's account applies to entire systems of the universe that satisfy certain criteria of simplicity and informativeness. While every consequence of the universe's optimal system is, as we discussed earlier, lawful, not every one of them is called a "law." Further, it is certainly not the case that the best system of the actual world entails "all planets travel in the same direction around the sun" since this holds only under certain contingent conditions. Given these conditions it is lawful that planets travel in elliptical orbits, but it is not a law.

However, some regularities may be so pervasive and central that they do qualify as laws. Of course, the fact that on the BSA a proposition that does not follow from accepted laws should be added as a fundamental law does not mean, as Roberts seems to suggest, that, as science develops, scientists should stop looking for explanations for it in terms of other laws and conditions. Whether or not the Past Hypotheses counts

---

[3] Roberts (2008) 23. Although Roberts is critical of the BSA his metaphysical view of laws is Humean since he thinks that laws organize the Humean mosaic, and we need them to find about it. He speaks of the universe being "law governed" but the sense in which he means "govern" is one that a non-Humean would consider deflated. It may be that at the end of the day when the details about what counts as scientifically best are worked out there is not much of a difference between the BSA and Robert's account.

as a law, according to the BSA, depends on whether it is entailed by the scientifically best systematization of the universe's Humean mosaic, and physicists do not currently and may never know what that system is. The only way to find out is to make alternative proposals, do experiments and calculations, and see which is best confirmed.

A second objection to the BSA is that the criteria for evaluating candidate systems may not determine a unique optimal systematization. What then does the BSA count as laws?

One response to this, due to Lewis, is that laws are the generalizations common to all the systems tied for best.[4] If the alternative best systems are not very different, this doesn't greatly conflict with the intuition that there is a unique best system. Another response is that there is no fact as to which system is best but that it is vague what the laws are. A third response is to relativize the notion of law to a system.[5] I prefer this response since it highlights the point that what makes a proposition lawful is its role in an optimal systematization and not governing. On the governing view there must be a unique set of laws that do the governing. The world can have only one governor. This rules out relativizing laws and vague laws. The systematization account allows for them.

A related, more troubling, worry is that the best systematization for our world may be so bad that it should not be counted as law determining at all. The BSA assumes that our world has a best system that is sufficiently good to count as law determining. One response to this worry would be to give up the concept of law, just as Nancy Cartwright suggested we should, once the theology she thinks is behind governing is rejected. Another response, of course, is to give up the BSA. But this pessimism is unwarranted and, in any case, premature. The history of science provides reason to believe that our world has systematizations that are quite good at satisfying the criteria that physicists propose for a system of the world, and that if there is more than one, that they will mostly agree on which generalizations are laws.

---

[4] "I doubt that our standards of simplicity would permit an infinite ascent to better and better systems; but if they do, we should say that a law must appear as a theorem in all sufficiently good true systems" Lewis (1973) 73.

[5] Massimi (forthcoming). "A Perspectivist Better Best System Account of Lawhood," in Ott and Patton (2018).

What makes Lewis's BSA Humean is that it accounts for laws and other nomological features of the world without appeal to fundamental necessary connections between property instantiations of distinct entities, or between governing or constraining laws and the events they allegedly govern or constrain. A way of understanding the BSA is that it promotes the systematizing aspect of the seventeenth-century concept of laws, while dispensing with the governing aspect. This does not mean that one cannot say that classical mechanical laws as understood by the BSA in some sense "govern" the motions of the planets. "Governing," for a Humean, consists in the fact that the motions of the planets satisfy the laws of classical mechanics. The laws needn't produce those motions. The fact that laws systematize is what makes them laws.

Later, we will see that the role that BSA laws play in counterfactuals also supports talk of their governing. Humeans can also make sense of talk of "powers," as long as powers are not fundamental but supervene on the Humean mosaic. On these Humean understandings, laws and powers don't add any metaphysical "oomph" to the universe but are a consequence of an optimal systematization of the Humean mosaic.

Mistaking the BSA for a conceptual analysis of the concept *law of nature* instead of a proposal for how we should understand what laws are is at the heart of some objections to it. This is especially the case for those objections that involve thought experiments which claim to show that there are distinct possible worlds with the same Humean mosaics but with different laws. If the BSA were an analysis, these objections would be devastating, since indeed there is an understanding of the concept of law under which they fail to supervene on the mosaic.[6] A simple example of such a thought experiment considers a world that contains only a single particle moving uniformly. The generalizations expressed by Newton's laws are true at this world, but the BSA would not count them as its laws. The best system for this world would just state that particles move uniformly. The argument is that since we can conceive that Newtonian laws are the laws of a one particle world, the BSA is wrong.

---

[6] Carroll (2010).

As Helen Beebee (2000) argues, this argument depends on a governance conception of laws or at least on a view of laws on which they are already understood as failing to supervene on the mosaic. When one is conceiving of the Newtonian regularities being laws in the one particle world, one may be imagining them as governing the world in a non-Humean sense of "governing." An advocate of the BSA grants that the concept of law of nature that scientists employ has both governance and systematizing aspects, but argues that, as a matter of fact, it is the systematizing that makes certain propositions laws and so we should be unmoved by the thought experiment. The Humean revised concept excludes the possibility of Newtonian laws being the laws of the one particle world, even though they are true at that world. More generally, on the BSA, and any other Humean account of laws, there will be worlds at which the propositions expressed by laws are true, but they are not laws. These worlds *undermine* the lawfulness of the actual world's laws, in other words, while these worlds are compatible with the content of those laws, they are not compatible with them being laws in those worlds. This is to be expected on the BSA since propositions are lawful because they are entailed by an optimal systematization of the actual world not of other worlds at which they may be true.

Physicists often do consider worlds that are models of laws, but whose best systematizations do not count them as laws. It is easy to see how the BSA can accommodate this practice by introducing the idea of laws relative to a source world. In our example, Newton's laws are laws of the one particle world relative to a Newtonian world.[7]

There are fancier thought experiments designed to show that laws don't supervene on a world's mosaic, but they all, like this one, depend on intuitions that already reject Humeanism, so I won't pursue this kind of objection further. The task of a philosophical account of laws isn't to agree with intuitions, but rather to show how laws according to that account are able to perform the jobs that scientists need laws to perform. Agreeing with philosophers' intuitions is not one of those jobs.

What are those jobs? The three most important jobs laws are called on to perform involve their roles with respect to counterfactuals,

---

[7] See Halpin (2003) where this approach is developed.

explanation, and induction/confirmation. An advocate of the BSA need to show that this account can perform these jobs at least as well, or better than, alternative accounts.

As we have discussed, propositions that are lawful support counterfactuals, but accidental propositions do not. A law supports a counterfactual if the law, together with other truths compatible with the law and "cotenable" with the counterfactual's antecedent entail the truth of the regularity.[8] For example, the law that bodies traveling elliptical orbits around the sun sweep out equal areas in equal times supports the counterfactual that if there were a planet orbiting the sun between Jupiter and Uranus it would sweep out equal areas in equal times. But the mere regularity that all coins in my pocket are dimes does not support the counterfactual that if this quarter were in my pocket it would be a dime.

Laws also play a special role in the evaluation of counterfactuals. The counterfactual "If there were a planet orbiting between Jupiter and Uranus, the laws of gravitation would still hold" is true.[9] Newtonian laws are also counterfactually resilient under counterfactual suppositions like this one. Counterfactual resiliency means that the law continues to be true under the supposition.[10] More generally there is something about lawful regularities in contrast to mere regularities that enables them to say what would happen under merely possible circumstances. The question is where does this ability come from? Non-Humeans attribute it to the necessity that laws provide regularities. It may seem, at first, that Humean BSA laws don't have the resources to enable them to support counterfactuals, or to enable counterfactual resiliency since they are mere regularities not backed by any necessity. But this is not correct. I will argue that these features come from the role of laws in systematizing and the way counterfactuals are evaluated.

---

[8] B is cotenable with A iff B is true and if A were true B would still be true. As Goodman pointed out, this makes the characterization of "cotenable" depend on a counterfactual, so as analysis, this is circular. How to avoid the circularity is an issue discussed in Loewer (forthcoming b).

[9] That is, laws are cotenable with any propositions that are consistent with the law.

[10] While resiliency requires that a lawful proposition continues to be true, that is not the same as its continuing to be lawful. No Humean account can satisfy that condition without exception.

This may be made clear by recalling that the most well-known and developed account of counterfactuals which is due to the arch-Humean David Lewis is compatible with Humean accounts of laws. Further, I will argue that given this account of counterfactuals, the BSA is able to explain why laws support counterfactuals and why they are counterfactually resilient for the most part.

According to the Lewis-Stalnaker similarity account, the counterfactual A > B is true iff in all the possible worlds that are most similar to the actual world at which A is true B is also true.[11] Lewis develops a particular version of a similarity account by characterizing similarity for relevant counterfactuals in terms of the extent of the regions of worlds that match the actual world with respect to fundamental events and the size and extent of conformity with respect to fundamental laws.[12] Overall counterfactual similarity is determined by balancing these two factors. Much more weight is given to minimizing the size of the regions in which the actual laws fail and the extent of this failure.[13]

Although on this account the truth conditions of counterfactuals are specified in terms of possible worlds, their truth makers are just non-nomological facts in the Humean mosaic. No fundamental necessity is involved. Reference to possible worlds is just a way of specifying the facts about the Humean mosaic that are truth makers and providing semantics for the logic of counterfactuals. Appeal to counterfactuals in this way doesn't presuppose Lewis's famous account of possible worlds

---

[11] Stalnaker's and Lewis's accounts differ in certain respects since Stalnaker requires a uniquely most similar world, and Lewis's account is a bit more complicated than my version here since he allows chains of similarity without a most similar world.

[12] There is much more about counterfactuals and Lewis's account in Loewer (forthcoming b). As will be shown, his particular account is defective, but that fact doesn't affect the point here that it is compatible with a Humean understanding of laws.

[13] Lewis's explicit recipe is that the factors determining the similarity ranking are: (i) big widespread violations of actual world laws; (ii) perfect match of fundamental fact; (iii) small local violations of actual world law; (iv) similarity of fact, in this order of importance. For example, "If the meteor that impacted off the coast of the Yucatan had missed the earth, dinosaurs would have continued to live for at least another 20 million years" is true iff in the most similar worlds to the actual world in which the meteor misses the earth the dinosaurs continue to live for another 20 million years. Lewis thinks that his account of similarity has the result that the most similar world to the actual world in which the antecedent is true are worlds that exactly match the actual world until a short time before the actual time of impact when the meteor swerves so as not to hit the earth. If the fundamental laws are deterministic, this will require a small violation of the fundamental laws. But after that violation the actual laws obtain.

as existing concrete entities of the same kind as the actual world. A possible world can be an abstract representation or a story about ways things can be. As far as this account of counterfactuals is concerned the laws can be Humean or non-Humean. If the laws are Humean, then the account of counterfactuals is Humean as well. A consequence is that a counterfactual proposition is just equivalent to a proposition that is *solely* about the Humean mosaic. In other words, a counterfactual proposition simply says that the actual Humean mosaic satisfies the conditions that make the counterfactual true. These conditions do not require the existence of *anything* other than the Humean mosaic.

On this account, laws support counterfactuals because of their role in characterizing similarity. But on Lewis's particular account, laws are not completely resilient under consistent antecedents.[14] The reason is that the most similar worlds to the actual world will, if determinism is true and the antecedent is false, be ones that include small violations of the actual laws. Lewis thinks that the violation occurs in a small region prior to the time of the antecedent, so it results in only a small compromise to resiliency. This is a special feature of Lewis's account of counterfactuals that is independent of his Humean account of laws. For now, I will employ Lewis's account of counterfactuals and argue that Humean BSA laws support counterfactuals as well as non-Humean laws do. Elsewhere I develop an improved account on which laws are resilient under nomologically possible antecedents and which is compatible with Lewis's Humean BSA.[15]

The question arises as to what it is about laws that make them so important in determining the similarity relation involved in evaluating counterfactuals. Non-Humeans can point to the fact that, according to their view, laws are among the fundamental ontological features of the

---

[14] This is the case when the laws are deterministic.
[15] In Loewer (forthcoming b), I describe a better Humean account of counterfactuals which dispenses with similarity in favor of probability, but on which laws both support counterfactuals and are resilient under most antecedents. Although on my account laws are completely resilient under counterfactual antecedents that are consistent with the laws, that they are laws may not be resilient under some counterfactual antecedents since on a Humean account like Lewis's there are propositions compatible with the laws on which the laws while true are not laws.

universe, as an answer to this question. The Humean BSA cannot say this but has, I think, a more informative answer.

The basic idea is described by Chris Dorst, who argues that the conditions that the BSA say make a regularity a law also show why laws are counterfactually resilient.[16] Dorst points out that the best systems for worlds like ours yield dynamical laws that enable us to use counterfactual reasoning to extend our knowledge of the world if we treat laws as especially important in evaluating counterfactual similarity. This is due to the fact that organizing information in terms of dynamical laws and initial conditions is scientifically optimal for both prediction and explanation. For example, to figure out where a meteor that hits the moon at a particular place was a day prior, one considers alternative hypotheses of the form "if the meteor had been at such and such a place a day earlier it would likely to have landed at such and such place today." We can then use Bayes's theorem or inference to the best explanation to infer where the meteor was yesterday. Evaluating counterfactuals like this one requires keeping the actual laws fixed.

A similar point concerns the use of counterfactual and subjunctive conditionals in rational decision making. When deliberating, we consider what would happen were we to make alternative decisions. The rational decision is the one that maximizes expected subjunctive counterfactual utility. This requires that laws are resilient when evaluating what happens in worlds that differ minimally from the actual world except for the decision. We know that all we immediately control are our decisions (which are events in our brain) and so we keep everything else at the time of making the decision fixed when evaluating what will happen. This includes the laws which, since they hold throughout time, can be used to provide an account of what will happen after alternative decisions.[17]

What enables us to extend our knowledge of a state at a time to subsequent times by counterfactual reasoning is that the laws are dynamical and can be applied to subsystems. Similarly, what enables

---

[16] Dorst (2020) provides a thorough account of the importance of counterfactuals in scientific reasoning and the role of dynamical laws in evaluating counterfactuals.

[17] In Loewer (forthcoming b), I provide an account of how this works in our world given its dynamical laws and probabilities.

counterfactuals to specify what will happen under alternative decisions is that the laws are dynamical and specify what follows from alternative actions in particular circumstances. Non-Humean dynamical laws would work as well. But it is the fact that laws are dynamical and apply to subsystems, not that they are non-Humean that accounts for their importance for evaluating counterfactual similarity. By emphasizing the organizing and systematizing role of laws, the BSA has an account of why it is that a scientifically optimal system will enforce a division between dynamical laws and initial conditions.[18] Some non-Humean views like Maudlin's just assume this. Others, like Armstrong's and Chen and Goldstein's, also provide no account. It is not clear how powers accounts support a division between states of systems and dynamical laws.

It must be granted that we have a very strong intuition that laws in some sense "govern." But I think this is due not to laws being over and above the Humean mosaic "pushing" it around or constraining it as non-Humean views maintain, but rather due to their role in counterfactuals. The counterfactual resiliency of laws accounts for the sense that laws constrain and govern. Whatever decision I make, or action I take, the laws would still obtain. No matter what I were to decide or how hard I were to try I would never travel faster than light. This gives the impression of laws being an outside force or restraint that prevents my traveling faster than light. But if the previous discussion is on the right track, the constraint is due to how we evaluate counterfactuals, not to the "oomph" supplied by governing laws.

While the sense in which Humean laws "govern" the mosaic will seem deflated to a non-Humean, it is, unlike non-Humean governing, not mysterious. The previous discussion shows that BSA laws do just as well as or better than non-Humean laws with respect to supporting counterfactuals. The more serious problem for the Humean BSA concerns the role of laws in explanations. It has been claimed that while non-Humean laws can perform this role, Humean laws, being mere regularities,

---

[18] Maudlin's governing view on which laws are FLOTES simply assumes that fundamental laws are dynamical while the other non-Humean views have no explanation for the centrality of dynamical laws. In contrast the BSA explains the centrality of dynamical laws.

cannot. Here is an argument by Tim Maudlin that clearly and forcefully makes this objection:

> If one is a Humean, then the Humean mosaic itself appears to admit of no further explanation. Since it is the ontological bedrock in terms of which all other existent things are to be explicated, none of these further things can really account for the structure of the Mosaic itself. This complaint has been long voiced, commonly as an objection to any Humean account of laws. If the laws are nothing but generic features of the Humean Mosaic, then there is a sense in which one cannot appeal to those very laws to explain the particular features of the Mosaic itself: the laws are what they are in virtue of the Mosaic rather than vice versa.[19]

The gist of the argument is that, on Humean accounts, the mosaic explains which propositions are laws, but since laws are supposed to explain their instances and these are included in the Humean mosaic, the mosaic explains part of itself. This allegedly results in vicious circularity, so something has gone wrong.

One feels the force of this objection if one thinks that the mosaic in some sense *produces* the law and then thinks of laws as explaining regularities by *producing* them. In this case, the Humean mosaic would be producing itself. But this misunderstands the relation between the mosaic and the laws. On the BSA, the mosaic doesn't produce the laws, and the laws don't produce parts of the mosaic.

My first response to the circularity objection was to call attention to the difference between scientific explanation and metaphysical explanation.[20] Humean laws *scientifically explain* events not by producing regularities but by unifying them. They accomplish this by subsuming them and being components of a system. That a regularity is a law is *metaphysically explained* by its being entailed by the best systematization of the Humean mosaic. As Elizabeth Miller (2015) points out, on the BSA the HM doesn't literally produce the laws either, even though it is true that the fact that the HM is a certain way explains why certain

---

[19] Maudlin (2007) 172.   [20] Loewer (2012).

regularities are laws. On the BSA, instead of the Humean mosaic *producing* laws, they are propositions that describe the HM. And, of course, BSA laws don't *produce* instances of the mosaic.[21] I argued that these observations remove the threat of circularity since nothing metaphysically explains itself and nothing scientifically explains itself. Instead, the law partially describes the Humean mosaic and scientifically explains events within it by unifying them.

Marc Lange (2013) replied to my response to the circularity objection by proposing a transitivity principle that he thinks connects scientific and metaphysical explanation and which shows that Humeanism implies that the Humean mosaic scientifically explains a part of itself after all. If this is correct, then the circularity problem reemerges. The principle is:

> If E scientifically explains [or helps to scientifically explain] F and D grounds [or helps to ground] E, then D scientifically explains [or helps to scientifically explain] F.

That is, when E helps to scientifically explain F, then that explaining is also being done by whatever D makes E the case. If D is what it is in virtue of which E holds, then D plays whatever roles in scientific explanations E is playing.

Lange provides some examples that conform to this transitivity principle. For example, increasing the temperature of a gas in a balloon scientifically explains the balloon's expanding and the increase in the motion of the molecules metaphysically explains why its temperature increases. It seems also to scientifically explain the expanding of the balloon.

Hicks and van Elswyk (2015) replied to Lange by disambiguating the transitivity principle and providing convincing counterexamples to the version that is required by Lange's argument. For example, the presence of a lion in a certain region may scientifically explain the number of prey animals in that region, but while the locations and motions of a few molecules that partly compose the lion may help to metaphysically

---

[21] Miller (2015).

explain the presence of the lion, they don't help to scientifically explain the number of prey animals. Had these molecules been elsewhere, the number of prey animals would have been the same. For Lange's argument based on his transitivity principle to have force, it would need to be exceptionless or at least he would need to show that it applies to the case of explanation by Humean laws. He has not established this and so it is not clear that the principle applies to it.

In any case, it is not difficult to see what is going on in anti-Humean arguments that Humean laws are incapable of supporting explanations. On non-Humean accounts, laws explain a regularity by producing it, or by resulting from the activities of powers that produce it. They explain by supplying "metaphysical oomph" that makes events occur. On both kinds of non-Humean accounts, a law involves something distinct from, but sufficient for its associated regularity. Although contemporary versions of these accounts of laws have dispensed with theological assumptions, the idea that laws explain regularities by producing them may be a remnant of the concept's theological origin according to which God supplied the oomph. In any case, Humean laws cannot explain in this way.

If Humean laws do not explain by governing or by producing, how do they explain? I have already suggested that they explain by systematizing and unifying. A system unifies disparate phenomena by showing how statements describing these phenomena are derivable from a few propositions or axioms that exhibit connections among the phenomena. For example, classical mechanics unifies the motions of pendula and cannon balls by showing how their trajectories derive from Newton's dynamical and gravitational laws. It similarly unifies terrestrial and celestial motions by implying that planetary orbits are elliptical. Showing that and how these phenomena follow the same laws explains them by exhibiting their connections.

A Humean best system also unifies counterfactuals since it specifies what happens were initial conditions to be different for many different initial conditions. For example, the fact that a rocket escapes the earth's gravitation field is explained by showing that its velocity exceeded escape velocity and that if its velocity had been appreciably lower it would have fallen back to earth. The fact that a law unifies events in the mosaic and that it is the entire mosaic that makes it a law does not mean that the entire mosaic unifies part of itself. This means that Lange's transitivity principle fails if laws are understood as explaining by unifying. A fuller account of explanation by unifying would specify exactly when a theory explains a body of events by unifying them and the degree to which it does that. I am not providing such an account here, but I think I have said enough to show how laws can explain by unifying and why this avoids the circularity objection to the BSA.[22]

BSA laws are involved in explanations in ways other than by unifying as well. Most important is that they support counterfactuals and causal relations which are involved in explanations. For example, $F = ma$ explains why a rock with a certain mass thrown with a particular force at a window causes the window to break by entailing the force with which the rock strikes the window.[23] The law supports the conditionals: "if the rock were not thrown, the window would not break" and "if the rock were to be thrown the window would break." As we have discussed, there are Humean accounts of causation in terms of counterfactuals and in terms of probabilities. If these accounts can be made to work, then the objection that Humean laws cannot play the role of laws in causal explanation completely loses its force.

It is often claimed that Humeanism about laws results in problems for inductive inference and confirmation that are not faced by non-Humean accounts. It has even sometimes been claimed that a non-Humean account of laws can enable a solution to Hume's famous problem of induction and rebut skepticism about induction. David Armstrong says, "if laws of nature are nothing but Humean uniformities, then inductive skepticism is inevitable."[24] He argues that such skepticism does not arise if laws are his governing laws. This is an amazing assertion. I will show that his arguments don't support it.

---

[22] Also see Friedman (1974) and Kitcher (1989) for accounts of how laws explain by unifying. Bhogal (2020) also rebuts the explanation argument against the BSA by arguing that scientific explanations unify while metaphysical explanations do not. Blanchard (2023) contains an excellent discussion of how BSA laws explain by unifying.
[23] A more complete explanation would also refer to the molecular structure of the window and the rock and the forces that maintain these structures.
[24] Armstrong (1980) 48.

Hume's problem of induction is often taken to be the problem of demonstrating, without begging the question, that it is rational to make certain inferences from evidence to propositions that are not logically implied by the evidence. One way of understanding what is required to show that the inference from E to A is rational is to show that if E is true then A is true or is likely to be true where "likely" is understood as an objective probability. If this is required to show that an inference is rational then Hume convincingly argued that this is not possible, since by assumption E does not logically imply A or that A is likely. Any additional premise that entails the truth or likely truth of A would itself need inductive support. This shows that a proposed justification of a system of inductive inference is inevitably circular.

Armstrong's approach to justifying induction is different. He claims that there is an a priori principle he calls "inference to the best explanation" (IBE) that makes certain inductive inferences rational and that it requires non-Humean laws. He argues that IBE supports the inference from the claim that observed Fs are Gs to the general claim that all Fs are Gs. First, a non-Humean law that Fs are Gs, i.e. N(F,G), is the best explanation for observed Fs being Gs. Second N(F,G) entails that all Fs are Gs. It follows that Fs so far observed to be Gs inductively supports that all Fs are Gs. In contrast, he claims, the existence of a Humean law that Fs are Gs does not explain why the Fs so far observed are Gs, and so inference to the best explanation cannot support the existence of the Humean law or the generalization it entails.

There are many problems with Armstrong's argument. First, it is not obvious that N(F,G) does explain why the Fs so far observed are Gs. The claim that N(F,G) explains why a particular F is a G is not like the explanation of why thunder explains the shattering of a window. In the latter case there is a mechanism involving the clap of thunder, the transmission of sound waves, the molecular structure of glass, and its breaking. This is a causal explanation. Showing that N(F,G) implies "all Fs are Gs" is the inference problem we discussed previously. But Armstrong just claims that N(F,G) explains why all Fs are Gs. Whatever this is, it is not a causal explanation. Further, in usual cases of scientific inferences via IBE, it is possible to independently test the explanans since positing it leads to further predictions. For example, when the existence of

Neptune was posited as the best explanation of perturbations observed in the orbit of Uranus, it lead to predictions about what telescopes would find. While we can test $(x)(Fx \rightarrow Gx)$ nothing like this seems in the offing for Armstrong's $N(F,G)$.

A third response to Armstrong's argument is that his claim that Humean accounts of laws cannot employ inference to the best explanation is not obviously correct. It may be, as he says, that a mere regularity does not explain its instances, so IBE doesn't by itself justify an inference from observed instances to the regularity. But what the Humean account does is instead explain why observed Fs are followed by Gs by appeal to the fact that it is a Humean BSA law that Fs are followed by Gs. Given our previous discussion, it is not obvious that the fact that a regularity is a Humean BSA law doesn't explain the regularity it entails. But in this case the explanation is not of the law causing the regularity, but rather it explains the regularity by appeal to a law that unifies it with other regularities.

There are some other objections to Humean accounts of laws related to Armstrong's and other arguments which also need to be discussed. There is the one I call "the Fluke Objection." According to it, if Humeanism is true, then the existence of best system laws should seem to be an accident or a fluke. We should not expect to find ourselves in a world which is systematizable, so regularities we do find should strike Humeans as flukes. The problem is not just that on Humeanism the existence of lawful regularities has no explanation, but that Humeanism undermines the reason to believe that simple universal regularities exist.

Here are John Foster and Galen Strawson making this argument against Humeanism:

> What is so surprising about the situation envisaged—the situation in which things have been gravitationally regular for no reason—is that there is a certain select group of types, such that these types collectively make up only a tiny portion of the range of possibilities, so that there is only a very low prior epistemic probability of things conforming to one of these types when outcomes are left to chance.[25]

---

[25] Foster (2004) 68.

One is presented with all these massy physical objects, out there in space-time, behaving in perfectly regular ways, and then one is told that there is, quite definitely, no reason at all for this regularity; absolutely nothing about the nature of reality which is the reason why it continues to be regular in the particular way in which it is regular, moment after moment, aeon after aeon. It is, in that clear sense, a pure fluke. It is, at every instant, and as a matter of objective fact, a pure fluke that the state of the world bears precisely the relation to the previous state of the world that one would expect, in line with the previous pattern of regularity.[26]

Humeans should grant that there is nothing in their metaphysics that *guarantees* or makes it likely that the world contains scientific regularities, or that it has a Lewisian best systematization, or explains why it does, if it does. They should agree that most Humean worlds are complex and unsystematizable. Within Lewis's metaphysics this follows from the fact that the recombination principle entails that the instances of perfectly natural properties can be distributed in space-time independently of each other and almost all the resulting worlds are unsystematizable.

But it doesn't follow that if there are BSA laws they are accidents or flukes. Strawson seems to think that Humeans are committed to fundamental properties being randomly distributed throughout mosaics and since systematizable mosaics are rare among mosaics it would be a "fluke" if our world happens to be one. But this is a mistake since randomness requires probability and there is no objective probability over Humean mosaics. On Lewis's account objective probabilities make sense only within a mosaic. They are specified by probabilistic laws that are entailed by the probabilistic laws of a best systematization of that mosaic. Given the best systematization of the actual world that includes a law specifying the half-life of radium, we can ask for the probability that a lump of radium will emit an alpha particle within a certain amount of time and even make sense of the probability of that mosaic given its laws. But it makes no sense to ask what is the probability that a

---

[26] Strawson (1989) 30.

mosaic is systematizable by laws that entail the probability that a radium atom emits an alpha particle or that an arbitrary mosaic has a best systematization at all.

Strawson is better off putting his objection in terms of what it is rational to believe if one believes Humean metaphysics. Strawson and Foster seem to think that if Humeanism is true, rationality requires a very low or 0 prior *subjective* probability in there being lawful regularities. This is explicitly the view argued for by Aldo Filomeno (2021). He appeals to the principle of indifference to support his claim. But one thing we know for sure is that, without limitation, the principle of indifference leads to contradictions. As far as I can see, no a priori principle of rationality dictates that Humeans should disbelieve that the world is systematizable or that it is unsystematizable. Indeed, if all one knows is all Humean mosaics are metaphysically possible and one then asked whether the actual mosaic is systematizable the appropriate response, is "I don't have a clue."[27] But given that we have found laws like those in quantum mechanics, general relativity, and statistical mechanics, it does seem rational to believe that the actual world is systematizable. Whether or not it is rational to believe a Humean account of laws is a completely different matter.

Dustin Lazarovici suggests that while it might be a mistake to claim that if Humeanism is true then it is unlikely (objectively or subjectively) that the actual Humean mosaic is systematizable, it is nevertheless the case that systematizable mosaics are atypical.[28] His reason is that there are uncountably many mosaics (every distribution of fundamental quantities at space-time points is a mosaic) but the number of mosaics that can be systematized in a scientific language is at most countable. Lazarovici thinks that this is a problem for Humeanism. His argument

---

[27] This is David Albert's response in conversation when asked what to believe when all one knows is that there are n possibilities one of which is actual.
[28] The notion of typicality has recently been invoked in statistical mechanics and other theories in physics in order to provide an account of explanation in such theories. For example, entropic behavior is typical among energetically isolated systems. A property or behavior is typical in a reference class if almost all members of the class have the property or behavior. Typicality is not a probability, but rather a "counting" notion. See Wilhelm (2019) for further discussion of typicality and Loewer (forthcoming) for a comparison of probability and typicality accounts of statistical mechanics.

invokes the principle that if a theory entails that an event is atypical, and the event occurs, then the theory incurs an explanatory deficit, which is a reason to reject it. This principle is a plausible principle when applied to scientific theories about a world, but it is less clear that it holds for metaphysical theories, which makes it a less than persuasive reason to reject Humean accounts of laws.

In any case, it is worth pointing out that the situation with respect to the unlikeliness or atypicality of the world having a best systematization seems, if anything, worse for certain non-Humean accounts of laws. On governing views, there is a possible world corresponding to every distribution of properties and every collection of governing laws, as long as the properties are distributed so that the laws are not violated. But just as systematizable worlds are a rarity and are atypical in the class of Humean worlds, they are also atypical in the class of these worlds. This class includes all the Humean worlds and in addition includes worlds with governing laws that are enormously complicated and gerrymandered. For example, there is a member of the class that is just like the actual world (assuming it is a Humean world) except it contains the governing law that all emerubies are gred. Aside from this, worlds with governing laws that cover many of their events are atypical. Unless the advocates of governing accounts add that such worlds are impossible, their view incurs the explanatory deficit of having no account of why our world, or their account of it is not like one of these typical worlds.

In the class of worlds whose fundamental properties are powers, systematizable worlds are rare as well since the powers interacting with one another can produce arbitrarily complicated patterns of events. Of course, it is possible to avoid this consequence if it is required that powers only combine to produce systematizable worlds, but this would clearly be a case of "theft over honest toil." Why should powers satisfy this condition?[29]

---

[29] The objection to Humeanism that it implies that it is a fluke for the actual world to be systematizable bears a resemblance to the "fine tuning of life" argument. The fine-tuning argument claims that possible worlds that contain laws with fundamental constants whose values are necessary for there to be life are atypical among all possible worlds. It goes on to argue that this provides reason to believe that a designer, i.e. God, who values life created the actual world with its laws and constants. The argument depends on comparing the probability of the laws supporting life given a designer with the laws supporting life given no designer and claims that

A second, somewhat different objection but just beneath the surface of the preceding remarks is that if the world's best system summarizes a Humean mosaic, it is puzzling that we can ever know or even have any rational beliefs about what the laws are. As Jenann Ismael has pointed out, since instantiations of perfectly natural properties are metaphysically independent of one another, whatever portion of the mosaic we have observed is metaphysically compatible with any distribution of perfectly natural properties whatsoever outside of that portion.[30] Examining the mosaic in region R or prior to time t can tell us nothing about the mosaic outside of R or after t if all we have to go on is the metaphysical structure of the mosaic. Ismael has shown how Humean metaphysics has led to Hume's problem of induction.

Humeans can respond by saying that anyone engaged in inductive reasoning, including those who hold non-Humean views of laws, must assume that the world is systematizable. Justifying induction is a problem for every account of the metaphysics of laws. Everyone must make the assumption that the world is systematizable to get induction off the ground. But the problem may seem worse for Humeans, at least those committed to Lewis's metaphysics, since, according to that metaphysics, systematizable worlds are a rarity.[31] I think Ismael has found a way of

---

the former probability is much greater than the latter. The problem with the argument is that there are no such objective conditional probabilities. Objective physical probabilities are specified by the world's laws, but those laws don't give probabilities for different laws and constants although they may give probabilities for different initial conditions. Objective epistemic probabilities are specified by a priori principles of rationality. But principles that have been suggested like the principle of indifference lead to contradictions and in any case don't supply the needed conditional probabilities. If the probabilities are subjective or epistemic, then anything goes, and the argument has no force. The fluke argument against Humean accounts of laws is defective for similar reasons. The laws of our world do provide objective probabilities for the set of the universe's initial conditions that result in life so one might attempt to reconstruct the argument using them. Discussing this would take us too far afield but I will return to it in Loewer (forthcoming b). In any case, this point doesn't provide any help for the fluke argument, since on Lewis's account the probability that the initial conditions of the world violate the Humean laws is negligible.

[30] Ismael (2023). Ismael argues that there is a problematic tension between Humeanism and the best system account of laws which is brought out by a conflict between Humeanism and the rationality of inductive inference beyond what we have so far discussed. I will argue later that her very interesting argument does not affect my PDA of laws that I develop in this book.

[31] While some non-Humean accounts of laws, like Maudlin's, and powers accounts don't commit to a view about metaphysical possibility, they also seem to endorse the claim that systematizable worlds are a rarity among possible worlds. If so, then the same issue affects them.

saying what really bothers many people about Humean accounts and their relation to induction. We will see that the view developed later in this book, while maintaining the idea that laws organize and systematize, helps with this problem and fares no worse than any other view with respect to the problem of induction.

So far as I can see, neither Humean nor anti-Humean accounts of laws and fundamental properties guarantees that the world is systematizable or contains any laws. There is no guarantee even on the governing account that there are any governing laws, but if it is accepted that they imply regularities then their existence does entail that the world contains corresponding regularities. If fundamental properties are powers, then they will be associated with regularities as long as powers don't conflict or compete. Without that, the world may not be systematizable. Humeans accept that the world may not be systematizable and hold the assumption that it is a posit of science. Neither kind of metaphysical account can justify induction. It is a mystery that the universe is lawful whether laws are Humean or non-Humean as much as it is a mystery why the universe exists at all.[32]

When the concept of mathematical laws of nature developed in the seventeenth and eighteenth centuries, the mystery was grounded in religion, and the faith that there are simple mathematical laws was based on faith in God. As God disappeared from the concept of laws, faith that the world is scientifically systematizable remained. Hume's argument shows that there can be no scientific justification for why it is systematizable if it is.[33] We can think of science, especially fundamental physics, as searching for systematizations and as an inquiry into the question of whether and to what extent the world is systematizable.[34]

---

[32] Those scientists of the seventeenth and eighteenth centuries—and philosophers like John Foster—who invoke God to explain why the universe has laws or why it exists may think they solve these mysteries but they only raise others. As Sidney Morgenbesser quipped in reply to the question "Why does anything exist?": "Even if nothing existed you would still complain."

[33] This doesn't preclude there being a scientific justification for why inductive reasoning works in a particular area in terms of more general scientific laws. For example, the inference from observed crows are black (albinos excepted) to all crows are black (albinos excepted) can be justified in terms of laws of genetics and evolution. Inferences to them can be justified by more general laws of physics. But this must stop at some point with fundamental laws. One might hope that inferences to them can be justified in terms of metaphysics, but I have argued against some such attempts to do that.

[34] I owe to David Albert this insightful way of putting the matter.

Some thinkers, such as John Foster mentioned earlier, think that in order to explain the fact, if it is a fact, that the world is systematizable by laws, one needs to invoke a deity as creator of the world who chose to create a world that is systematizable. A similar thought lies behind much discussed fine-tuning arguments that attempt to show that the particular values of fundamental constants provide evidence for a creator. But whatever it is, the hypothesis of a creator is not a *scientific* hypothesis. There is no way of testing it and no mechanism posited for how the creator operates. As far as physics is concerned, it seems more appropriate to have faith that the world is systematizable by laws then to hypothesize a creator responsible for laws.[35] Science depends on faith a much as religion does.[36]

Here is where we have gotten to so far. The idea that the primary goal of physics is the discovery of laws of nature and that laws are the engines of scientific explanation emerged during the seventeenth century. At that time laws of nature were conceived of as principles that describe how God governs the motions of matter or as instruments He employs to enforce those principles. The principles were thought to exhibit characteristics worthy of the deity; they are mathematical, simple, elegant, comprehensive, eternal, deterministic, and so on. They systematize and organize his creation.[37] Contemporary views about laws, Humean and non-Humean, do not presuppose theology. They are perfectly coherent without God. But just because they don't involve God, more needs to be said about the nature of governing and constraining. Over time, the theological presuppositions of this view of laws fell away but the view that laws both govern and form a system of principles that satisfy these characteristics persisted. As the sciences developed, the requirements that a law determined system should satisfy were developed and refined. Contemporary metaphysics of laws has divided into two camps: the Humean BSA and the non-Humean governing and powers accounts.

---

[35] Of course, this is not an argument against there being a creator, just against confusing theology with science.
[36] The contents of these faiths are very different as are the consequences of having them. This is not the place to further discuss and evaluate them.
[37] Marc Lange pointed out to me that the idea that Humean laws systematize the world owes as much to the theological origin of the concept as does the non-Humean view that laws govern.

The first deflates the governing aspect and promotes the systematizing and organizing aspect, and the second elevates the governing aspect while not emphasizing the systematizing and organizing aspect. Both approaches can be developed into coherent metaphysical accounts of laws. However, we have seen that appealing to governing or powers involves metaphysical commitments which replace the role of God that work, as He is said to, in ways that are just as mysterious. I have argued that the Humean account dispenses with these mysteries, and that many of the objections to it rely on remnants of the theological origin of the concept of law. I have argued that the scientific work that laws actually perform in science has no need of governance or powers but relies on the systematizing and organizing aspects. I think the best way to view the debate concerning the metaphysics of laws is as between alternative proposals for replacing the theological account of laws that emerged during the eighteenth century. While I don't think that I have shown that non-Humean views are mistaken and the Humean BSA is correct, I do think that enough has been said to establish the BSA as a serious contender. However, Lewis's particular version of the BSA itself has some very serious problems to contend with. I will next discuss those problems and then describe my new account, the PDA, and how it deals with them.

5
Problems with Perfectly Natural Properties

Up to this point, I have developed and defended Lewis's Best System Account of laws, and argued that while it does not capture all the intuitions that philosophers and others have about laws, it does a good job in accounting for the roles of laws in physics.[1] I now want to consider two big problems with the account. The first is that, by relying on the perfectly natural properties, it threatens to make its laws irrelevant to physics. The second is that, by relying on the criteria of simplicity and informativeness, fit, and others derived from the history and practice of physics in evaluating candidates for best system, it threatens to make laws and chances dangerously mind dependent. This problem Lewis calls the threat of "ratbag idealism." I address perfectly natural properties in this chapter, and postpone discussing ratbag idealism until after introducing the PDA, which dispenses with Lewis's ontology of perfectly natural properties, and for which ratbag idealism may seem to pose an even worse problem.

Lewis's BSA is built on the assumption that there is an elite class of properties he calls "perfectly natural." The idea that some properties are more "natural" than others, and that some are maximally natural, has a long history in philosophy, but Lewis's account is most immediately connected to Armstrong's theory of universals (1978). Lewis's account of properties differs from Armstrong's, but they both hold that degrees of naturalness and being perfectly natural are objective and metaphysically

---

[1] I am grateful to participants in my graduate seminar at Rutgers spring 2023 and especially to Ted Sider for discussion of the issues in this chapter.

fundamental.[2] Most philosophers agree that some properties are more "natural" than others, in that some properties are more suitable for appearing in laws, explanations, and inductive inferences than others. What makes Lewis's and Armstrong's view particular is that they hold that naturalness is a feature the properties possess independently of contingent facts about a world and independently of human interests, beliefs, and practices.

Examples of natural properties and degrees of naturalness that are usually given are: green is more natural than grue, and mass and spin are more natural than green and may even be perfectly natural. But what makes a property perfectly natural, or one property more natural than another? What features characterize these properties?

Lewis mentions several conditions that perfectly natural properties satisfy and jobs he recruits them to perform:

(i) They are a unique small subset of all properties, i.e. are sparse.[3]
(ii) Which properties are perfectly natural is an objective matter of metaphysics independent of human interests.
(iii) The distribution of their instantiations throughout a world's space-time is a supervenience base for all truths at that world.
(iv) They are categorical and intrinsic to the points or individuals at which they are instantiated.[4]
(v) They are instantiated at space-time points or concrete point-sized individuals.
(vi) They are the basis of an account of degree of naturalness for all properties and an account of objective similarity.
(vii) Every possible world is a possible space-time arena decorated with instantiations of perfectly natural properties and every

---

[2] Armstrong holds that properties are universals that are instantiated by actual individuals and exist only if instantiated, whereas Lewis holds that properties are classes of possible individuals.
[3] For Lewis, a property is a set of possible individuals (or space-time points), and every such set is a property. An n-place relation is a set of ordered n-tuples of possible individuals.
[4] A property is categorical if it is individuated independently of nomological and causal relations, and necessary connections with properties instantiated at distinct points. This allows for necessary connections between properties instantiated at the same point.

space-time arena and every such a decoration is a possible world.
(viii) Not every perfectly natural property is instantiated at the actual world. The properties not instantiated at a world w he calls "alien" to that world.[5]
(ix) They are reference magnets that play a central role in accounts of the reference of mental and linguistic representations.[6] Degrees of naturalness play a role in determining reference by selecting references of predicates from among candidates that satisfy other criteria in terms of their naturalness.
(x) One of the aims of fundamental physics is the discovery of perfectly natural properties. Possible examples of perfectly natural properties are mass, charge, and spin.
(xi) Thay are the referents of predicates in languages in which candidates for a world's scientifically optimal systematization are formulated. Laws are theorems of the true systematization and so refer to perfectly natural properties.
(xii) They are the basis for solving Goodman's "New Riddle of Confirmation," since perfectly natural properties are projectible and the more natural a property, the more projectible it is.

Lewis seems to grant that some relations are more natural than others and suggests that the only perfectly natural relations instantiated at the actual world are geometrical relations, but seems to allow that physics may discover others. Lewis calls the claim that the only actually instantiated perfectly natural relations are geometrical, and that perfectly natural properties are instantiated at points, and that all truths supervene on their space-time distribution "Humean Supervenience" (HS). He doesn't claim that HS is true but wants to investigate whether it may be true.

---

[5] Lewis's account of possible worlds is based on perfectly natural properties being categorical, and a recombination principle, according to which every combination of instantiations of perfectly natural properties instantiated at different point is metaphysically possible.

[6] Putnam argued that the truth of a theory of the world doesn't determine the reference of the predicates of the theory since there are many distinct mappings of predicates onto references that make the theory true. Saying that perfectly natural properties are reference magnets means that naturalness is involved in constraining mappings.

It is not clear which of the entries on the list above Lewis considers to be constitutive of what it is to be a perfectly natural property, and which are not constitutive but jobs they are recruited to perform. The roles of perfectly natural properties in Lewis's theories of laws, modality, reference, similarity, and confirmation may not be constitutive of what it is to be perfectly natural, but rather they are part of an account of what laws are. If that is so then if these accounts fail, then it would not show that there are no perfectly natural properties. For example, if HS fails, or if charge and mass turn out to not be perfectly natural, Lewis's metaphysics of perfectly natural properties would survive. The only entries on the list that seem certainly constitutive are that perfectly natural properties are sparse, that their distribution of their instances at a world form a supervenience basis for all truths at that world, that it is a matter of objective metaphysics which they are, and perhaps that they are categorical and intrinsic. But if this is all that fixes what it is for a property to be perfectly natural then that is not much guidance, since Lewis holds that every set of possible individuals (or space-time points is a property) there are many different sets of properties that satisfy these few conditions. On the other hand, if it is required that natural properties satisfy all the entries on the list, or even most of them, then there is a worry that, if nothing satisfies them, there are no perfectly natural properties.[7]

Lewis thought that discovering which properties are perfectly natural is one of the tasks of physics. He provides as examples of what may be perfectly natural properties mass and charge. But it is important to be clear that Lewis does not say that a requirement on a property's being perfectly natural is that physics, even ideal physics, says that they are fundamental. Even ideal physics may fail to identify perfectly natural

---

[7] It is interesting to observe that requiring perfectly natural properties to play roles in Lewis's accounts of laws, modality, similarity, reference, and confirmation keeps these accounts from being empty. We saw that emptiness results in the BSA without a restriction on the languages in which candidates for best system are formulated. Goodman's "grue paradox" shows that such a restriction is needed in accounts of confirmation and Putnam's paradox of reference shows it is needed by accounts of reference and without such a restriction everything turns out to be similar to everything else. But it is not clear that the restriction to perfectly natural properties or even the same restriction is needed for each of these theories.

properties although it attempts to identify them.[8] This is a version of metaphysical realism about fundamental perfectly natural properties.

The main problem with Lewis's BSA is that it is built on the metaphysics of perfectly natural properties. This makes its account of scientific law beholden to metaphysics. In my view, while metaphysics should depend, as much as possible, on science and especially physics, science should depend as little as possible on metaphysics.[9] I take it that one of the main, perhaps *the* main, job of metaphysics is to provide an account of what reality must be or might be like for science to be successful, or as successful as it appears to be.[10] I will argue that physics does not depend on there being a unique set of metaphysically prior perfectly natural properties satisfying all the features Lewis attributes to them.

The first problem with basing the BSA on the metaphysics of perfectly natural properties is that contemporary physics posits fundamental entities, properties, and relations that are not compatible with HS and seem to violate some of the conditions Lewis places on perfectly natural properties. Quantum theory, in particular, creates difficulties for the claim that all fundamental properties are categorical and intrinsic and especially for HS's additional claim that the only relations are geometrical spatio-temporal relations since QM seem to include in its fundamental ontology whatever is represented by the wave function and whatever that is it is not intrinsic to points in three-dimensional space. Lewis was worried about quantum mechanics:

> I am not ready to take lessons in ontology from quantum physics as it now is. First, I must see how it looks when it is purified of instrumental frivolity...of double thinking deviant logic...and—most of all—when

[8] Thanks to Veronica Gomez for pressing this point.
[9] This is not to say that all there is to metaphysics is science. One of the tasks of metaphysics is to investigate the presuppositions of science. Lewis seems to think that the existence of perfectly natural properties is one such presupposition. I will argue that it is not.
[10] Metaphysics also aims, in Wilferd Sellars's memorable words, to understand "how things in the broadest possible sense of the term hang together in the broadest possible sense of the term." My view of metaphysics is similar to that espoused by Ladyman and Ross in *Everything Must Go* (2007), but as will be clear I don't share their ontic structural realism or views about modality. Seeing how everything hangs together may involve the kind of conceptual analysis engaged in by analytic metaphysicians that they disdain.

it is purified of supernatural tales about the power of the observant mind to make things jump.[11]

That may have been a reasonable attitude to take in the 1980s, but by now there are several versions of quantum theory that have been purified of instrumental frivolity, deviant logic, and consciousness making things jump, and which have explicit ontologies and laws. All these accounts violate HS in one way or another. Most important is that it is difficult to see how HS can accommodate the quantum mechanical state since it cannot be represented as occupying 3+1-dimensional space-time. On some views, it can be thought of as a kind of field occupying a high dimensional configuration space. The field values do not seem to be Lewis's perfectly natural properties or quantities since they are not independent of one another.[12] On other views, the quantum state is more like a law or a component of a non-Humean law that does not exist in space-time but "governs" the motions of particles in space-time. This view is clearly incompatible with Humeanism.[13]

Even fields in classical physics seem to be in tension with the claim that fundamental properties are instantiated at points and are categorical. If the electromagnetic field is a fundamental entity, it is present throughout all of space-time; if its energy is a fundamental property, as it usually is considered to be, then it is a global property of the field not instantiated at points. One could imagine a defender of Lewis's metaphysics maintaining that it is field values that are fundamental and that

---

[11] Lewis (1986) xi.
[12] This account has been advocated by David Albert (1992), (2000), (2023) and Alyssa Ney (2013). See Loewer (1996) for a way of saving the letter of HS by positing that the fundamental physical space is 3n dimensional configuration space and the QM state is a field in it. Michael Esfeld, Eddy Chen, Harjit Bhogal, Zee Perry, and others have proposed versions of Bohmian mechanics that are compatible with HS but at the cost of excluding the wave function from fundamental ontology. On their account, the fundamental ontology consists solely of point particles. Their idea is to construe the quantum state not as part of the ontology but as systematizing the trajectories of particles. That is, more along the lines of the way that Lewis understands laws. This approach is compatible with HS but at the risk of adding a possibly a very complicated description to the best systematization.
[13] This is a view advocated by Goldstein and Maudlin. Michael Esfeld's super-Humean Quantum Mechanics attempts to treat the quantum state as a kind of Humean law but faces various difficulties. I discuss problems with Esfeld's views in the next chapter. A Humean version of Eddy Chen's Wentaculus account of quantum theory together with statistical mechanics may overcome these problems.

these are instantiated at points and that the energy of the field is supervenient on the field values, but this seems to be putting the interpretation of a physical theory into a metaphysical straight jacket.

Further conflicts with HS come from quantum gravity theories that posit entities that are not point-like (e.g. strings and branes), space-times that have more than 3 + 1 dimensions, and theories in which space-time is not even fundamental but is claimed to emerge from something more fundamental.[14] So, there is reason to think that HS is not true or, at least that physicists do not feel constrained by it. Since Lewis understood HS to be a conjecture that might turn out to be false, he might not grieve if it is given up. The Humean mosaic can be expanded to include properties that are instantiated by regions (or individuals) that are larger than point size (like strings) and relations other than geometrical relations (like entanglement relations). But it is not clear that this expansion is compatible with perfectly natural properties being categorical and their role in the construction of possible worlds.

There are further developments in contemporary physics that call the entire Humean ontological basis on which Lewis's BSA is built into question. As mentioned some current physical theories posit fundamental properties and quantities that are not intrinsic to the regions in which they are instantiated and are not categorical and whose ontologies do not fit easily into the framework of individuals and properties.[15] It has been claimed that this is the case for quantum field theory and the standard model of elementary particles. For example, French and McKenzie argue that QFT and the standard model of elementary particles contain symmetry principles that individuate certain fundamental quantities and entities, and so imply that the properties and entities instantiated in distinct regions are not intrinsic and involve necessary connections which are:

> …the properties through which the fundamental constituents of matter interact in terms of gauge transformations, and these bring in their

---

[14] Wüthrich and Lam (2023).
[15] One of the examples that Lewis gives of a perfectly natural property is spin but on every version of quantum mechanics spin is not an intrinsic but is a contextual property. Exactly what value it has depends on how it is measured.

wake the appropriate gauge bosons. Then it looks as if we have no choice but to say that the properties such as charge and color are not the sort of properties that lone objects can have, and hence that these properties are not after all intrinsic.[16]

Another example is that the symmetry principles of the standard model that individuate elementary particles entail that it is metaphysically impossible for there to be fermions with their particular masses unless there is a Higgs field. This seems to make mass a non-intrinsic property, so it doesn't qualify as perfectly natural. Tim Maudlin goes further, suggesting that it is not right to think of QFT on the model of entities bearing properties and relations at all. According to Maudlin, the QFT representation of certain quantities (e.g. quark color) as fiber bundles implies that these quantities are not metaphysically independent:

> We should note that adopting the metaphysics of fiber bundles invalidates a set of modal intuitions that have been wielded by David Lewis under the rubric of the Principle of Recombination. According to Lewis, Hume taught us that the existence of any item puts no metaphysical constraints on what can exist adjacent to it in space. This invites a cut-and-paste approach to generating metaphysical possibilities: any object could in principle be duplicated elsewhere, immediately adjacent to the duplicate of any other item (or another duplicate of itself). Duplication is supposed to be a metaphysically pure internal relation between items. But from the point of view of fiber bundle theory, it makes no sense to "copy" the state of one region of space-time elsewhere even in the same space-time, much less in a disjoint space-time. There is no metaphysical copying relation such as the Principle of Recombination presupposes.[17]

One could try to save the letter of Lewis's Humeanism in the face of these examples by adding further fundamental relations, and understanding space-time as higher dimensional, thus further violating HS. But an alternative would be to admit that a true fundamental theory

---

[16] French and McKenzie (2017) 53.   [17] See Maudlin (2007) 103.

may posit an ontology that is not representable in terms of a Humean mosaic, while seeing if it is possible to save the core idea of the BSA that what makes a proposition express a law is not governing, but its role in systematizing.

Most of the preceding problems could be due not to perfectly natural properties themselves, but to Lewis's theory of modality being based on recombination and the requirement that perfectly natural properties are categorical.[18] If these are removed from the list of conditions constitutive of perfect naturalness, then arguments that physics is incompatible with Lewis's naturalness lose much of their force. All that would be left from our list of conditions on perfectly natural properties is that they are sparse, and the distribution instances of a set of perfectly natural properties form a supervenience base for each world. But since there are many sets of properties that could serve as a supervenience base for all truths, we have lost our grip on which properties are the perfectly natural ones. What distinguishes the perfectly natural properties instantiated at the actual world from all the other sets of properties that are instantiated and serve as a supervenience base of actual truths?

A different, although related, problem with Lewis's reliance on natural properties as the basis of the BSA is that physics doesn't seem to care about which properties are perfectly natural. This is argued for by Schafer and Hicks. They claim that Lewis's BSA seems to endorse a principle they call "Link" that is out of step with practice in physics:[19]

Link: Only predicates referring to perfectly natural properties can be invoked in candidates for the best system.

Link says that the language in which the scientifically best system is formulated contains in addition to mathematical and logical expressions (including a probability function) only expressions that refer to perfectly natural properties. Although Lewis thinks that it is the job of physics to discover the perfectly natural properties, it is not clear that physics has any interest in discovering properties that satisfy all of

---

[18] This is Ted Sider's view (personal comm.).   [19] Hicks and Shaffer (2017).

Lewis's conditions on perfect naturalness.[20] As we have discussed, the laws found in fundamental physics often are expressed in terms of predicates that do not refer to properties that satisfy all the conditions on perfect naturalness. Being an electromagnetic field, acceleration, force, energy don't seem to qualify as perfectly natural properties, although perhaps Lewis would say they can be defined in terms of expressions referring to perfectly natural properties. Physicists often take theories to be equivalent even though the theories consider different properties to be fundamental. I don't mean that these theories are just empirically equivalent but that they apparently say the same about physical reality, except for what is fundamental. For example, classical mechanics can be formulated in a number of different ways—Newtonian, Hamiltonian, Lagrangian—that have the same laws but seem to count different properties and quantities as fundamental. At most one formulation can satisfy Link. Physicists may prefer one formulation to another for particular purposes on certain occasions, but not because they think that one is formulated in terms of expressions referring to the truly fundamental properties/magnitudes. Quantum mechanics provides another example of a theory which has different formulations that are generally regarded as equivalent by physicists.[21] The bottom line is that while physicists are interested in discovering fundamental laws, it is not obvious that they care about discovering a unique set of fundamental properties. But that means they are not interested in perfectly natural properties since perfectly natural properties are supposed to be uniquely fundamental.

An issue that has received much discussion is whether categorical properties are or have unknowable quiddities. Quiddities are for properties what haecceities are for individuals—something (we know not what) that makes it the property it is independently of its causal and

---

[20] Lewis wants perfectly natural properties to play central roles in his accounts of modality and reference, but it is not obvious that the properties that a fundamental theory in physics deems fundamental should play either of these roles.

[21] I don't mean merely empirically equivalent but as describing the same physical world even though they count different properties as fundamental. See North and Coffey for discussions of whether these formulations should be counted as the same or different theories. This seems to be a problem for Maudlin's account as well since his account supposes that there is a unique fundamental law that governs the evolution of states.

nomological relations. This makes them unknowable since it is possible for distinct properties to have the same nomological profiles. A consequence is that it is possible for categorical properties to swap causal and nomological roles in different possible worlds or even within a single world without any way of our knowing, since we only know a property by way of its causal effects. So, for example, if negative charge and positive charge are perfectly natural properties then there are two possible worlds which have the same nomological structures but in which instantiations of positive and negative charge switch places. It is even possible that within a single possible world there are distinct categorical properties that play indiscernible nomological roles. For example, there is a world in which there are ten different perfectly natural properties all playing the role of negative charge. Perhaps there are quiddities that play the roles of charge and mass on even-numbered days and switch to play the roles of mass and charge on odd-numbered days. If there are, physics is completely ignorant of them and doesn't care about them.

Such metaphysical possibilities lead Lewis and Langton to endorse a view they call "Ramseyan humility," the view that there are facts which are irremediably unknowable since there is no way to discover which world we are in. I am not sure that Ramseyan humility really does follow or, if it does, that this is a problem. My point is that quiddities apparently make no difference of importance to physics. They are not what physicists think they are referring to in their fundamental theories. For this reason, it would be better for scientific metaphysics to avoid them if possible.

Quiddities can be avoided by supposing that fundamental properties are not categorical but possess nomological roles that individuate them. This may seem plausible since when physicists posit particles and their properties it is together with laws that involve them. It is difficult to think that a property could be mass if it doesn't play the nomological role of mass or that anything that does play that role could be anything other than mass. One version of this view called "nomic structuralism" claims that a property just is a node in a web of nomological relations. Nomic structuralism avoids positing scientifically meaningless metaphysical distinctions, but it has struck some philosophers as unsatisfactory since it seems to imply that all there is to reality is structure without content.[22] Russell referred to this view with the cryptic remark:

> There are many possible ways of turning some things hitherto regarded as "real" into mere laws concerning the other things. Obviously, there must be a limit to this process, or else all the things in the world will merely be each other's washing.[23]

Russell seems to be saying that there must be some fundamental properties that are not individuated by laws, otherwise no property is genuinely individuated. Which properties are the ones not individuated by their roles in laws? Even space-time geometrical relations seem to be individuated in terms of nomological relations to matter and energy in general relativity.

We are left with a puzzle. It seems unsatisfactory that all fundamental properties are categorical as Lewis's account sems to require, but it is also unsatisfactory that all fundamental properties are individuated solely by the laws to which they conform. Perhaps some properties are one and some the other and some a combination of a quiddity together with essential laws. But we don't possess a principle that tells us which is which. It would be better for our metaphysics of laws and properties to be non-committal about this issue and leave it to physics to decide, if it needs to, which fundamental properties are categorical, which are fundamental dispositions, and which are mixtures of the two, or, if physics fails to do that, to dispense with the distinction altogether.[24]

Bas van Fraassen (1989) raised a different problem which also seems to show that physics doesn't care about Lewis's perfectly natural

---

[22] Ladyman and Ross (2007) hold a related view they call "ontic structural realism." One challenge the view faces is to account for what makes a structure physically realized. Related to this is that the same structure seems to be realizable in different ways. For example, there seem to be worlds in which electrons and positrons and other such particle pairs switch roles.

[23] The view that Russell is criticizing is that all properties are individuated entirely in terms of their necessary connection to one another.

[24] Michael Esfeld (2019) has developed a view he calls "super-Humeanism" in which the only fundamental property is being a particle and the only fundamental relation is distance. All other predicates and function terms in a theory are understood as devices for systematizing particle trajectories in much the same way that the BSA treats probabilities. This avoids issues re quiddity swapping. I discuss super-Humeanism later in detail.

properties. Van Fraassen pointed out that it seems possible for there to be a true theory Σ that optimally satisfies all the criteria implicit in physics, i.e. fulfills all of Steven Weinberg's dreams for a "theory of everything" (TOE),[25] and yet not be a Lewisian best system since its fundamental predicates fail to refer to Lewis's perfectly natural properties. If so, the fundamental properties referred to in the TOE will not be perfectly natural and the class of propositions it claims express laws may not match the class deemed laws by a world's Lewisian BSA. Van Fraassen's point is that while physics aims to find nature's scientific joints, these may not coincide with her Lewisian metaphysical joints, assuming she has them. This can result in a mismatch between the laws of the BSA and the laws according to ideal physics. The fact that physics is not interested in satisfying Link suggests that this can come about. Here is an example. On Lewis's account, there is a possible world in which different perfectly natural properties (i.e. properties with different quiddities) play the nomic role that we think mass plays in our world (assuming our world is Newtonian) in different space-time regions. If so, then the language in which the best system is formulated will contain different predicates for each of these "mass" properties and one of the axioms of the system will be a disjunction specifying the behavior of each of these mass properties. It would thus be more complicated than the best systematization formulated in the language of physics which counts these properties as identical, but it still may agree on those propositions physics counts as laws. It could turn out that the system Σ that results by dropping some of the disjuncts associated with regions with few particles will be sufficiently simpler to overcome the decrease in informativeness. If this is the case, then the laws specified by that system will specify the Lewisian laws while the physicist's theory will be the usual Newtonian theory even though mass is not a perfectly natural property. Call this "the mismatch problem."[26]

---

[25] Weinberg (2011). I am assuming that among these criteria are that the theory is not merely empirically adequate but also true. However, I don't assume that the theory says which properties are perfectly natural or even which are fundamental.
[26] The mismatch problem is discussed by van Fraassen (1989); Loewer (2007); and Demarest (2017) who suggested its name.

A defender of Lewis's view of the role of perfectly natural properties might respond that the mismatch problem is not a problem at all but simply a consequence of realism about laws and properties and no different from other skeptical worries. She might claim that there are metaphysical facts about which properties cut nature at its metaphysical joints and even a theory that satisfies all the other scientific desiderata may be mistaken about which properties mark the joints and which true propositions are laws. But it seems to me that this takes realism too far. It is presumptuous for a metaphysician to say to a physicist who believes she has found a true TOE that optimally satisfies all scientific criteria that she may not have discovered the fundamental scientific properties and laws. Those who follow the program of naturalizing metaphysics should say that it is physics and the other sciences, not metaphysics, which determines nature's joints and reject this response. If physics does not determine a unique set of joints, then we should conclude that nature does not have unique joints.

Another response is to claim that the mismatch problem doesn't arise due to Lewis's account of reference magnetism. On that account, a true TOE's fundamental predicates must refer to perfectly natural properties since they are attracted to them as if by a magnet. The properties that the predicates refer to are those perfectly natural properties that make the theory true. This is ingenious, but Lewis's account of reference is contentious.[27]

Why should the basic predicates in a TOE refer to Lewisian perfectly natural properties? Lewis doesn't answer this question, but he does propose his account of reference magnetism to natural properties as a response to what has come to be known as "Putnam's paradox." I will return to discuss this in Chapter 10.

---

[27] Lewis suggested something along these lines in a conversation I had with him at a metaphysics conference in Syracuse shortly before he died. But we couldn't see how it can be made to work since it seems that different natural properties can satisfy the theory. Cian Dorr remarks that " with true scientific theories stated in a language whose basic predicates were too unnatural for the axioms to meet the bar for lawhood would have to be quite remote and unusual"; Dorr (2019 sec. 2.6). But reference magnetism is controversial, and, in any case, it is not clear that it will avoid the problem.

This is related to but distinct from the mismatch Shamik Dasgupta calls "the problem of missing value."[28] The problem is that if perfect naturalness is a metaphysical primitive there needs to be an explanation of why physicists or anyone other than metaphysicians who believe in them should value finding perfectly natural properties and laws characterized in terms of them. A TOE that optimally satisfies all scientific criteria but fails to match the Lewisian best system because it is not formulated in the language of perfectly natural properties does not seem to lack any value of interest to science. In fact, satisfaction of Lewis's requirement that the law determining optimal system is formulated in the language of perfectly natural properties may come at the cost of satisfying some of the criteria that science does value. The problem is that it is not clear that requiring the account of laws to be based on perfectly natural properties adds any value.

Dasgupta observes that there are many collections of properties whose instantiations can serve as the supervenience bases for all truths. Instead of formulating the best system in terms of systematizing all the facts specified in the language of perfectly natural properties, we could formulate best systems in terms of a language whose primitive predicates refer to the perfectly natural* or perfectly natural** properties and so on. Such systems may determine different "laws"—laws*, laws**, and so on. The question is "what is it about perfectly natural properties and laws characterized in terms of them that makes them suitable to guide physical theorizing and determine which propositions are those that physics seeks to find and call laws?" It is not adequate to simply claim that "they are the ones that guide physical theorizing because they are perfectly natural" or because they cut nature "at the joints." This would be, to borrow Lewis's quip, like saying that someone has strong arms just because he is called "Armstrong." Why does a property deserve to be called "perfectly natural" or "a joint of nature?"

---

[28] Dasgupta (2018) 288. Dasgupta's discussion is detailed and subtle. I understand him as providing an argument that an adequate account of laws should explain what it is about laws as characterized by Lewis that makes discovering them objectively valuable for scientists.

Lewis assumes that reality comes equipped with perfectly natural properties and that it is the job of physics to find them, and laws characterized in terms of them. It is supposed to be an objective matter what the optimal system and the laws are and that physicists ought to try to find them. The problem is that without an explanation of why natural properties have normative force, we have no more reason to think that the best theory formulated in the language of perfectly natural properties rather than the best theory formulated in the language of perfectly natural* or natural** properties determines the laws that physicists aim to find. Dasgupta's challenge is to provide an account of why Lewis's perfectly natural properties possess value. He claims it cannot be met without compromising the objectivity of natural properties.[29]

If we had a solution to the mismatch problem, and could show that discovering or attempting to discover a TOE is objectively valuable, we would also have a solution to the problem of missing value.[30] We could then say that Lewis's perfectly natural properties derive their value from the value of laws determined by the best systematization. But we have no argument that the properties referred to by the fundamental predicates occurring in the scientifically optimal theory coincide with Lewis's metaphysical posits of perfectly natural properties. Dasgupta argues that accounting for the value possessed by our division of the world into scientifically fundamental properties is to understand them as *relative* to us and our values. He suggests that this leads to the kind of anti-realism and relativism associated with Nelson Goodman. Even if it doesn't go that far it seems to exacerbate the worry that Lewis's has that the BSA succumbs to "ratbag idealism." I will later argue in Chapter 10 that the PDA provides an account on which the properties that physics counts as fundamental are valuable without sliding either into pernicious

---

[29] Dasgupta observes that it is ironic that Lewis raises exactly the missing value problem to accounts of objective probabilities which take them to be metaphysically fundamental propensities. He writes "Objective probabilities are supposed to be credence guiding via Lewis's Principal Principle and its ilk, but Lewis thinks that there is no explanation of why a metaphysically primitive propensity should guide credence. I just paraphrased Lewis's famous objection to anti-Humean conceptions of objective chance because my objection to realism is exactly analogous. Ironic, then, that the problem with realism can be found in the writings of someone I take to be an arch realist! But ironies aside, let us review Lewis's argument so as to use it as a guide."

[30] I owe this observation to David Albert in conversation.

relativism or ratbag idealism and thus avoids the problems of "missing value" and "mismatch." It does this without relying on positing properties that are metaphysically prior to properties posited in physical theories.

One final point is that, while I have been focusing on perfect naturalness, Lewis's account also includes claims that one property is more natural than another and claims about degrees of naturalness. For example, according to Lewis, the property *green*, while not perfectly natural, is more natural than the property *grue*. Degrees of naturalness might play a role in accounting for special science laws and the reference of non-fundamental expressions and other matters. Lewis doesn't develop such accounts and he doesn't provide much detail about degrees of naturalness although he proposes that the degree of naturalness of a property is determined by how expressions referring to it can be defined in terms of perfectly natural properties.[31]

To recap, the problems with tying the BSA to perfectly natural properties are:

(i) There are properties considered to be fundamental in theories proposed in physics that don't satisfy all of Lewis's conditions of perfect naturalness.
(ii) Fundamental physics, while interested in laws, doesn't seem interested in finding a unique set of fundamental properties. Equivalent scientifically best systems can be formulated which take different properties to be fundamental.
(iii) Perfectly natural properties are quiddities and so are unknowable by science but there are fundamental properties in physical theories that seem to have at least some of their nomological roles necessarily.
(iv) It leads to the mismatch problem.
(v) And it leads to the missing value problem.

All these problems point in the same direction. Physicists don't care which properties are metaphysically perfectly natural and physics has no need for them.

---

[31] See Lewis (1983) and Dorr (2019) for more discussion of degrees of naturalness.

Lewis's makes the metaphysical proposal that there is a class of perfectly natural properties that are the supervenience base for all truths and play roles in theories of modality, reference, similarity, and confirmation. He thinks it is the job of physics to discover them. A different view is that physics develops criteria for what counts as an optimal theory of the world and the fundamental properties physics posits should count as the ones we call "perfectly" natural whether or not they play all the roles that Lewis thinks perfectly natural properties play. This is a kind of Euthyphro question. Remember that Socrates asked Euthyphro whether an act is virtuous because the gods love it or whether the gods love it because it is virtuous. Similarly, are properties scientifically fundamental because they are perfectly natural or are they perfectly natural because they are scientifically fundamental? Lewis's account endorses the first answer while the PDA, which will shortly be described, endorses the second.

6
Super-Humeanism

Before describing the PDA, I want to examine another attempt to dispense with metaphysically fundamental perfectly natural properties that also preserves Lewis's idea that laws are determined by a scientifically optimal systematization. By getting rid of dependence on most perfectly natural properties it seems, at first, to avoid most of the problems I just discussed. This approach, called "super-Humeanism," has been advanced and developed especially by Michael Esfeld.[1] Its basic idea is to claim that the fundamental metaphysical ontology of the world consists solely of propertyless particles that persist for all time and bear changing distance relations to one another. There are no fundamental properties or relations other than distance relations. The distance relations constitute spatial relations among particles from which the dimensionality and geometry of space is constructed. They are required to satisfy the triangle inequality and connectivity condition so that every particle is connected to every other particle by a distance relation at a time. However, there is no requirement that the distance relations conform to those of a three-dimensional (or any dimensional) space or conform to any particular geometry. It thus differs from Lewis's BSA which takes space-time points in a three-dimensional Euclidian space as basic individuals and constructs perduring individuals out of properties instantiated on continuous space-time trajectories.[2] Where Lewis **seems** to assume absolute space-time (though I don't think it is essential to his account), super-Humeanism endorses a relationist account.

Esfeld also assumes that change of distance relations is fundamental. This seems to mean that he considers time to be fundamentally different

---

[1] Esfeld and Deckert (2017).
[2] Sometimes Lewis appears to prefer an ontology of point particles and posit a relational account of space time with respect to space.

Laws of Nature and Chances: What Breathes Fire into the Equations? Barry Loewer, Oxford University Press.
© Barry Loewer 2024. DOI: 10.1093/oso/9780198907695.003.0006

from space. By requiring that trajectories of individuals are continuous, super-Humeanism rejects Lewis's view that there are no necessary connections between events in non-overlapping regions. Also, by treating time to be special in this way, it differs from the BSA which can construe time as one of the four dimensions characterized in terms of patterns of property instantiations.[3]

According to super-Humeanism, particles on trajectories comprise the fundamental ontology of the universe. Although Esfeld says that the particles are propertyless, what he means is that they have no *fundamental* properties other than distance relations at times. Particles and sums of particles may have other properties, but these are reducible to classes of particle trajectories. This means not just that particles can be identified with a class of particle trajectories but that what properties a particle has at world w supervenes on the totality of particle trajectories at w. On this view the world, or what we can think of as the super-Humean mosaic, consists entirely of particle trajectories. A scientific theory attempts to systematize the super-Humean mosaic by specifying a relatively simple axiom system Σ that provides information about it. The best systematization is the one (or one of those) that optimally balances information and simplicity. Laws are generalizations entailed by the best system. Super-Humeanism is like Lewis's BSA shorn of perfectly natural properties and left only with particle trajectories as its fundamental ontology.

Esfeld's idea is to treat predicates that seem to refer to fundamental properties as devices introduced to systematize particle trajectories. The language in which a candidate systematization is formulated may include predicates satisfied by certain particles but not by others. In this way, the system can provide information about particle trajectories. "Particle Q has mass = x" means only "Q satisfies the predicate has mass = x" is entailed by the best systematization of the super-Humean mosaic. These predicates don't refer to fundamental properties of

---

[3] Understanding the arena in which fundamental properties/magnitudes are instantiated as an n-dimensional space-time enables one to formulate Lewis's account so it is compatible with Einstein's relativistic theories. It is not obvious how metaphysical accounts that treat the temporal dimensions as metaphysically fundamental can do the same. Lewis's doesn't endorse this view or say very much about time, but this kind of account is compatible with the BSA.

particles but rather they classify them to assist in formulating a simple and informative systematization of their trajectories. Predicates introduced in this way, play roles that are similar to the role that probability plays in Lewis's BSA. Just as there is no fundamental property *probability*, there are no fundamental properties such as *mass*, but additional predicates play a role in laws that provide information about the super-Humean mosaic.

Super-Humeanism takes items that are usually thought to be part of ontology, and places them in the realm of the nomic. Esfeld suggests doing the same to whatever the quantum mechanical wave function is or represents. This is an intriguing suggestion. There are many different accounts of what, if anything, the wave function represents.[4] These range from an epistemic state to something occupying Hilbert space that as it evolves determines all facts in the world and in many worlds, to something similar to a non-Humean law along the lines of Maudlin's account of laws, to a complex valued field.[5] Since the quantum mechanical state doesn't occupy three-dimensional space, but configuration space, this last view involves positing that the fundamental space is high dimensional configuration space, and the field occupies and evolves in that space.[6] All of these proposals encounter problems and puzzles. But if Esfeld's super-Humeanism works then the wave function can be understood as a device for systematizing particle trajectories and so there is no need for a high dimensional fundamental space or the other exotic ontologies. But I will argue that it doesn't work.

Esfeld's proposal is bold and interesting. He doesn't claim that it is a priori that particles are the metaphysically fundamental ontology of the universe, but he does claim that a fundamental particle ontology is sufficient to interpret any scientific theory that has so far been devised. His claim is plausible since particle ontology has the feature that it is not difficult to discern at least the material portion of the manifest image of

---

[4] See Albert 1992 and Maudlin (2019) for alternative accounts of what the quantum mechanical wave function represents.

[5] The idea that the wave function represents an epistemic state is quantum Bayesianism; see Fuchs et al. (2014). It treats the wave function in a way that is similar to epistemic accounts of objective probability.

[6] Albert (1992); Ney (2013).

the world in it and since any theory of the world must account for and is tested by the positions and motions of particles it is prima facie plausible. Many of the properties that we think of macroscopic objects possessing can be understood in terms of their motions or dispositions to motion. The properties that seem unlike these, for example color or the taste of lemon, seem to be or be related to phenomenal consciousness and may be beyond the purview of physical theory.

Super-Humeanism dispenses with Lewis's perfectly natural properties and doing so avoids problems they bring with them. It avoids the problems of quiddities and so the worry that there are properties that we cannot know and that there are objects that have the same dispositions but instantiate different perfectly natural properties. It does this not by claiming that fundamental properties are dispositional but by rejecting that there are fundamental properties other than particle position. So, super-Humeanism avoids the dilemma of choosing between fundamental properties being categorical or dispositional. It also avoids the mismatch and missing value problems because there are no perfectly natural properties for its fundamental predicates to fail to match or questions about the value of matching them. Obviously, it is valuable to discover the positions of material objects and the particles that compose them. At first, super-Humeanism looks pretty good. However, as we will see, it encounters problems and difficulties that render it unacceptable.

Exactly what does the best systematization of the super-Humean mosaic look like? Esfeld is not completely explicit about this. There seem to be two alternatives. I will describe them for a Newtonian world, but the same point holds for other worlds.

The first alternative is an axiom that assigns particular values to the terms "mass" and "charge" for each particle, so that the Newtonian laws, $F = ma$, and the force laws are satisfied. The second doesn't assign values to mass and charge to each particle but instead contains an axiom that says that there exist values that can be assigned to the mass and charge of each particle so that Newtonian laws are satisfied. The first axiom system is enormously complicated for a universe like ours since it will involve assigning values of mass and charge to an enormous number of particles. Even if there are only a few kinds of particles where all particles of a given kind have the same mass, charge, etc., the axiom will be enormously complicated since it will require identifying the kind of each particle. This makes it unsuitable as a candidate for the best system.

The second alternative will be much simpler but at the cost of resulting in radically indeterministic dynamics. By that I mean that given this axiom, the fundamental intrinsic state of a system at a particular time and the laws do not entail the intrinsic state at other times. The axioms and the positions of particles at t doesn't entail where the particles are at other times.[7] This axiom system is too uninformative to count as a good candidate for the best system. What this shows is that by excluding certain properties from the fundamental ontology super-Humeanism has undermined its account of laws.

Vera Matarese raised a second problem.[8] She pointed out that since super-Humeanism places no restrictions on the predicates that candidates for best system can introduce to systematize the super-Humean mosaic, it allows adding a predicate $F(x)$ that holds of a particle x iff x belongs to the super-Humean mosaic being systematized. The axiom $ExF(x)$ then would entail all the propositions true of this and only this Humean mosaic, and it is very simple. If informativeness is measured as Lewis proposed in terms of possibilities excluded, then $ExFx$ optimally systematizes the mosaic. The result is a system Σ that is very simple, but it also has the disastrous consequence that all general propositions true of the mosaic are laws.

It was a problem like this that led Lewis to restrict the languages for formulating candidate systems to those whose predicates refer to perfectly natural properties in the first place (he also later allowed the introduction of a probability function that does not refer to a perfectly natural property). But Esfeld can't require this or anything else that involves positing fundamental properties over and above particle positions.

---

[7] The laws of this system will look exactly like the usual Newtonian laws and so may appear to be deterministic. But since mass is not an intrinsic fundamental property of particles at a time, they are not genuinely deterministic. Even if one knows the complete fundamental state of the world around the small interval of time t and the laws of this world, one would not be able to infer the state at other times. This point is made by David Albert (2023) in the context of arguing against attempts to remove the wave function (quantum state) from a theory's ontology and place it in its nomology.

[8] Matarese (2020a) and (2020b).

A possible response is to reject F(x) simply because it trivializes super-Humeanism, but not only does this seem ad hoc, it won't work since there are many other concocted predicates that lead to similar unacceptable results. Super-Humeanism needs an objective way of selecting which predicates can be added to systematizing languages so that the best systems determine what physicists would count as laws. In Lewis's account this is accomplished by specifying that the predicates refer to perfectly natural properties. Matarese's argument shows that until such a way is identified, super-Humeanism can't be considered to provide an adequate account of laws.

Even if super-Humeanism can find a way of dealing with Matarese's and my objections without introducing further ontology, the fact that its ontology is restricted to propertyless particles creates further problems. One of the primary goals of scientific theorizing is to produce explanations. For example, classical mechanics explains why two objects under the influence of a mutually attractive force, assuming they are isolated from other influences, will accelerate towards each other with accelerations that are inversely proportional to their masses. It is in virtue of instantiating a particular value of mass that an object moves on a particular trajectory. But super-Humeanism cannot understand explanation in this way, since according to it, for an object to have a certain mass just is for it to move on certain trajectories. In general, super-Humeanism is incapable of endorsing a causal explanation of the motion of any object, since the properties whose instantiations we would normally cite as causes of the motion supervene on the totality of trajectories of objects including the motion of this object. Any such proposed explanation appears to be defectively circular.

In defense of super-Humeanism, Esfeld argues that it is no worse off than views that construe properties like mass as powers to produce certain motions. These also seem to be circular. Even if this is correct, the point does not apply to perfectly natural properties, since they are individuated independently of powers and laws and the trajectories they explain.

A second reply is that super-Humeanism does to mass what the BSA has already done to laws so should not be objectionable to a Humean about laws. Just as the BSA understands laws as not producing regularities but rather to result from a scientific systematization of the Humean mosaic, super-Humeanism takes mass and other properties not as producing causes but as aiding in the systematization of trajectories of particles. I think this is correct, but there is a difference between the two cases. It is mysterious how laws can produce events and it is easy to see how they systematize and unify them. This is one of the main points in favor of the BSA. But it is usual to understand properties like mass as causes of motion. Super-Humeanism is asking for a radical revision in how we think about properties and explanation.

If the preceding objections are not enough, the cost of eliminating all fundamental properties except particle positions is proves to be even higher when we consider its consequences for counterfactuals and indeterministic laws. Suppose we ask what would happen were certain particles located differently from their actual locations at a time t. We should then look for trajectories that are similar to the actual trajectories.[9] Since there are no perfectly natural properties, the particles on these new trajectories cannot be required to possess the same or similar perfectly natural properties. There is no reason why these alternative trajectories need to assign the same mass values to the particles. But if it is required to assume that the trajectories also conform to the actual laws then the resulting trajectories may turn out to be ones whose best systematizations assign to them different masses. In other words, had the particles been located differently their masses might have differed. I don't think we want this counterfactual to come out true.

The situation is even worse if the fundamental laws are indeterministic. In this case the actual Esfeld mosaic is best systematized by indeterministic laws that assign probabilities to alternative evolutions of a trajectory at a time. Some of those alternative trajectories are best systematized by assigning very different laws and masses to particles. In other words, given the positions of all the particles at t there may be positive probabilities of their having different masses at later times.

---

[9] In Loewer (forthcoming b), I propose an account of counterfactuals that dispenses with similarity in favor of probability, but the same objection to super-Humeanism holds for that account.

These counterfactual positions undermine particle masses. Call this "super undermining." It is bizarre!

The BSA also permits "undermining" futures. These result in histories whose laws and probabilities though compatible with the actual laws and probabilities are incompatible with these being the laws and probabilities. This may strike one as counterintuitive, but it is not difficult to see that it doesn't lead to any real difficulties for the actual world. The situation for super-Humeanism with respect to properties like mass is much more troubling. Lazarovici concludes that super-Humeanism is a "starving ontology."[10] I agree, so despite super-Humeanism's initial attractions we need to look elsewhere for an adequate account of laws and fundamental ontology.

---

[10] Lazarovici observes that since a super-Human ontology consists solely of propertyless particles with changing distance relations, it seems an amazing coincidence that the arrangement of these distance relations satisfy a particular geometry. This is similar to his objection to the BSA that systematizable worlds are atypical. But it seems much worse for super-Humeanism than for the BSA, since the latter posits a three-dimensional space as part of its fundamental ontology; Lazarovici (2018).

7
The PDA

We now come to the Package Deal Account (PDA) which I will argue is an account of laws, chance, fundamental ontology, and fundamental space-time, that emphasizes the systematizing role of laws as do the BSA and super-Humeanism, but rejects the metaphysics of both and so, I will claim, avoids their problems. The PDA preserves the core Humean idea that what makes a generalization a law is its role in systematization and unification but dispenses with Humean metaphysical commitments. I will argue that it is superior to all the accounts we have so far canvassed.

Lewis also describes his BSA as a "package deal." His reason is that it explains:

> …why the scientific investigation of laws and of properties is a package deal; why physicists posit natural properties such as the quark colors in order to posit the laws in which those properties figure, so that laws and natural properties get discovered together.[1]

What Lewis says is correct, but his BSA package is methodological not metaphysical. Even though Lewis says that proposals for laws are posited together with proposals for perfectly natural properties and the space-times in which they are instantiated, on the BSA the properties and space-time are metaphysically fundamental and the laws derivative on the distribution of the instantiations of perfectly natural properties throughout space-time. In contrast, as I will explain, the PDA is a package in which fundamental properties, space-time, and laws are metaphysically on a par. None is metaphysically prior to the others. Further, in the PDA, the package includes non-fundamental special science properties and laws, and candidates for a fundamental theory is

[1] Lewis (1983) 368.

Laws of Nature and Chances: What Breathes Fire into the Equations? Barry Loewer, Oxford University Press.
© Barry Loewer 2024. DOI: 10.1093/oso/9780198907695.003.0007

formulated in a language sufficient to express them. By making these changes to the BSA, the PDA extends a best system account of laws to ontologies and accounts of space and time, or space-time relations, posited in contemporary physics, that are not committed to Lewis's metaphysical presuppositions. It thus avoids many of the problems encountered by Lewis's Humean BSA.

Unlike super-Humeanism, the PDA doesn't assume a one-size-fits-all fundamental ontology of propertyless particles, but rather leaves the ontology and fundamental properties up to physics. However, it accepts super-Humeanism's (and for that matter, Descartes's, John Bell's,[2] and others') claim that the first job of physics is to account for the positions and motions of material bodies and how these positions record the measurements of other quantities. By building on this, while rejecting both fundamental ontologies of perfectly natural properties and propertyless particles as the basis for an account of laws, the PDA moves in the direction of naturalizing metaphysics. This way, it moves toward a view of fundamental properties and laws from inside the perspective of science, in contrast to Lewis's and Esfeld's more a priori "God's eye," and metaphysics-first perspective.

Unlike Lewis's BSA, the PDA doesn't assume a given four-dimensional fundamental space-time structure, or any space-time structure at all.[3] Nor does it assume, as Esfeld does, that time is a special dimension Instead of an absolute space-time, the PDA allows but doesn't require theories formulated in terms of relational accounts of space-time along the lines of Mach and Barbour.[4] It also allows for theories that posit many dimensional space-time structures, and to some contemporary quantum gravity theories that view space-time not as fundamental but as emergent from something else.[5]

The question that immediately arises is "if the PDA doesn't assume a basis of fundamental perfectly natural properties, or propertyless particles instantiated in space-time, what do candidates for optimal system

---

[2] Bell (1987).
[3] The absolute three-dimensional Euclidean space + time is likely not essential to Lewis's account, but he never says how he would extend the BSA to cover alternatives.
[4] Mach (1960); Barbour (1982).   [5] Wüthrich and Lam (2023).

systematize?" Without an answer to this question, what prevents the PDA from emptiness or succumbing to the trivialization that befalls Lewis's BSA without perfectly natural properties? What is to prevent the PDA from selecting "gruesome" properties and ontology as fundamental, or even Lewis's Fx true of all and only actually existing individuals as a fundamental property and ExFx as the best theory? Answering this question involves a brief discussion of the nature and goals of physics.

In my view, the best way of understanding the enterprise of physics is that it begins, as Quine says, "in the middle," with the investigation of the motions of macroscopic material objects, for example, planets, projectiles, pendula, pointers, and so on, as described in the languages employed by scientists.[6] Physics advances by proposing and testing theories that include laws that cover and explain the motions of macroscopic objects and whatever ontology and quantities are introduced along the way.

To account for the motions of macroscopic material objects, classical mechanics introduced an ontology of point particles and quantities like mass, charge, force, energy, and so on. Laws relating these quantities also systematize and explain the behaviors of macroscopic objects. To account for interactions between light and matter, Faraday and Maxwell proposed that there are electromagnetic forces and fields and laws describing them. To account for the melting of ice, the spreading of smoke, the growth of plants, and other thermodynamic phenomena, it was proposed that macroscopic objects and systems are composed of molecules in constant motion. To explain the chemical composition and the behavior of molecules and their macroscopic effects, it was posited that they are themselves composed of atoms which are composed of charged particles that produce electromagnetic fields. Accounting for the stability of atoms and the ways they combine to form molecules and interact with electromagnetic radiation involved the development of quantum theory. These theoretical developments not only systematized phenomena introduced by earlier theories, but also advanced systematization of the macroscopic phenomena with which the process began.

---

[6] See Quine (1981) and Verhaegh (2018). David Albert suggested the aptness of Quine's way of characterizing physics and its relation to the PDA.

In a number of instances, as in the case of classical mechanics and Maxwellian electromagnetics, conflicts arise whose resolutions lead to the introduction of further theories with new accounts of space-time and ontology and laws, as was the case for relativity.[7] The primary lesson of this whirlwind history of physics is that in the course of seeking to provide laws that systematize and explain the motions of the material objects with which the process began, physics introduced further ontologies and properties, space-times to locate them, and laws that systematize the whole package.

The ultimate, perhaps unrealizable, goal of this process is the discovery of what Steven Weinberg calls "a theory of everything" (TOE) that specifies a fundamental space-time, fundamental ontology and properties, and fundamental laws that cover not only the motions of macroscopic objects with which physics begins, but also whatever additional ontology and quantities have been introduced along the way. A TOE is a true theory that systematizes its fundamental ontology together with the motions of macroscopic objects and satisfies, to the greatest extent possible, the criteria that physicists have developed for evaluating fundamental theories, i.e. truth, simplicity, informativeness, fit, providing explanations, and so on. Among these criteria truth is crucial. As physics develops, missteps may be made and theories which are not true are proposed, but as physics progresses, they are revised and replaced. Physics aims for a *true* theory of everything. A theory that meets all the criteria except truth won't do as an account of fundamental laws, properties and spacetime. A theory of everything that meets all the physicists' criteria except perhaps truth may be false but it is the best that can be justifiably proposed as a system that determines laws and fundamental properties.

After truth, the most important of the criteria for evaluating proposed TOEs is that the fundamental ontology and properties of a candidate TOE provide an explanatory supervenience basis for macroscopic objects and their motions and whatever non-fundamental properties and laws have been introduced as science develops. It is the motions of

---

[7] Examples of this are the development of special relativity to resolve conflicts between classical mechanics and electromagnetic theory and the development of quantum mechanics to account for the stability of atoms.

material objects that provide observational and experimental evidence for evaluating proposals about the laws and ontologies introduced to explain them. This process requires that the ontology of an adequate best system and its laws provide a supervenience base not only for macroscopic objects and properties but also for macroscopic laws, causation, and counterfactuals.[8] This means that the distribution of its fundamental ontology and properties will also be a supervenience base for all objects and properties that can causally interact with ordinary macroscopic objects that are covered by laws.[9]

Physicists' usual understanding of a TOE doesn't explicitly include an account of how macroscopic facts are made true or false by fundamental facts. But a package that systematizes fundamental ontology together with macroscopic ontology, like the PDA, needs to include an account of how the fundamental and non-fundamental are connected. Proposals for fundamental physical theories are not only tested by observations of macroscopic events but aim to provide an explanatory basis for all phenomena whether observable or not.

It has become usual to call the connection between the more and less fundamental "grounding" and say that fundamental facts ground non-fundamental facts. This makes it sound like grounding is a worldly relation connecting distinct entities.[10] But I don't think this is the right way to think of grounding. Examples of connections between macroscopic

---

[8] For this to be the case David Albert and I have argued that the fundamental theory needs to include a probability distribution over physically possible histories of the universe. We call the package of fundamental dynamical laws and the probability distribution "the Mentaculus" and argue that special science laws, counterfactuals, and probabilities are all derivable from it. See Albert (2000) and (2015); Loewer (2021). These issues are discussed in detail in Loewer (forthcoming b).

[9] I am characterizing a fundamental system or TOE as aiming to specify a fundamental ontology that provides a supervenience base for all physical properties. This is short of what is usually understood as physicalism, since it allows for epiphenomenal properties that relate to physical properties by bridge laws that do not supervene on physical laws and facts.

[10] Jonathan Schaffer (2017) calls such principles "metaphysical laws." Schaffer says that these metaphysical laws explain how higher-level facts and properties are *grounded* in more fundamental facts and properties. I agree, but he and I are thinking of the connections quite differently. For him they are on a par with laws of physics that connect the state at one time with the state at other times. He even thinks of them as similar to causal laws and grounding similar to causation, though they differ in that metaphysical laws and grounding relations hold with metaphysical necessity. But in the PDA, the principles that connect the macro and the micro are very different from the laws of physics. They are based on the fact that fundamental terms acquire meaning via their being introduced as part of a theory that explains macroscopic expressions.

and more fundamental entities are "macroscopic objects are composed of the particles where they are located," "the mass of a macroscopic object is the sum of the masses of the particles where it is located," "the temperature of a macroscopic system is the average kinetic energy of the particles that compose it." These connections are not laws like the laws of physics or of the special sciences, since they are not contingent. But they are not exactly analytic and although they are constrained by meanings, they are not a priori. They play the dual role of characterizing the nature of the fundamental ontology and properties in terms of antecedently understood macroscopic ontology and properties and showing how the macroscopic facts supervene on fundamental facts. For example, the concept "atom" is introduced as referring to very small entities with mass and energy located where, for example, a rock is located, so that certain macroscopic features of the rock (its total mass, location, and motion) are determined by the properties of the atoms together with the laws. This characterization of atoms explains how other types of material objects and their properties supervene on the distributions of atoms. One doesn't need a special grounding law or a whole collection of unrelated "metaphysical laws" connecting atoms with macroscopic objects. The characterization of what an atom is—a small material constituent of paradigm material objects like rocks—together with functional characterizations of macroscopic properties and the laws is sufficient to *lock down* the connections between the fundamental ontology and the macroscopic ontology. Giving a complete and detailed account of this would be an immense undertaking but it is pretty clear that it is in principle possible at least for ordinary macroscopic objects and properties.[11]

The answer to the question of what the PDA says that an optimal system aims to systematize is this. Candidates for an optimal system of the world aim to systematize truths about the motions of macroscopic objects expressed in the languages with which science begins together

---

[11] A much-discussed exception to this account is phenomenal concepts and properties. Following Descartes, Chalmers used the failure to account for phenomenal properties in term of physical ones along the lines I just described to argue for dualism about consciousness. Dualism doesn't follow, since this failure derives from the very special nature of the concepts we apply to our own phenomenal states; see Loar (1997); Balog (2012).

with whatever truths considered to be scientifically relevant as science develops to achieve its goals of systematization, explanation, and prediction.

The languages with which science begins are shaped by human biology, psychology and history. They include terms for types of macroscopic bodies and their manifest properties, for example shapes, colors, motions, etc. Although these languages can express "gruesome" ontologies and properties, science is not aiming to account for phenomena described in such vocabularies although it can do so. Science aims for a theory that enables relatively simple accounts of the motions of pointers, pendula, projectiles, planets, etc. described in natural and scientific languages. But it doesn't care how complicated its accounts of arbitrary mereological sums of particles may turn out to be. Similarly, science aims at theories that enable simple accounts of changes in shapes and colors of typical macroscopic objects as characterized in ordinary language with which science begins. This requirement penalizes adding "grueish" expressions as a theory develops. This makes science an enterprise that is to an extent relative to human biology, psychology, and history. But that is as it should be since science is a human invention. Further, I submit that given what we know about our world, it is plausible that any intelligent beings that engage in constructing a scientific system of the world that accounts for the motions of macroscopic objects will end up with pretty much the same accounts of laws and fundamental ontology. If they aim for a true, simple, informative, and explanatory theory, whatever their psychology, biology, and starting language, they will converge on similar systems of the world. I will return to this issue in the chapter on realism and relativism when discussing the threat of "ratbag idealism."

Quine's characterization of science as a web of propositions connected by explanation and prediction is epistemological. On his account, experience impinges on the periphery and may lead to adjustments throughout the web even at its inner-most core. The goal is the production of a true theory of the world that is capable of explaining all physical phenomena. The PDA is a metaphysical counterpart to Quine's epistemological account of science. A system satisfying the PDA's criteria specifies which truths express fundamental laws and which terms refer to fundamental ontology and quantities. Recall that these criteria include that the theory is true, that the truths expressed in its fundamental vocabulary are a supervenience base for all physical truths, that it maximizes informativeness, explanatory effectiveness, and simplicity to the extent possible, and so on.

Macroscopic concepts are conceptually and epistemologically prior, but their referents are not physically fundamental. On the other hand, the concepts referring to theoretical entities and properties introduced in accounts of macroscopic phenomena are derivative of macroscopic concepts. So, there is a kind of circle from macroscopic concepts to fundamental concepts and from fundamental ontology and properties to macroscopic ontology and properties. It is not a vicious circle but a virtuous one. Fundamental concepts depend epistemologically and conceptually on macroscopic concepts and macroscopic objects and properties depend metaphysically on fundamental objects and properties.

These connections play a central role in determining the reference of theoretical concepts and our understanding of them. Some examples are that the understanding of microscopic particles is dependent on our understanding them as components of macroscopic objects located where the objects they compose are located and the understanding of the quantum mechanical state is dependent on its connection with the motions of macroscopic objects, for example pointers, cats, etc.

David Albert calls the relation between functional concepts/properties and the theoretical concepts and ontologies that account for them "anchored functionalism." His idea is that functionally characterized macro properties are anchored in fundamental ontology, properties, and laws in the sense that we understand the latter in terms of the former. For example, a functionally characterized rock, as something with a certain causal profile is anchored in particles that compose it, and particles are *understood* as point objects with mass that compose macro-objects like rocks. This last is the important point. We understand what particles are by the role they play in accounting for macro objects like rocks.[12]

---

[12] According to Albert, this plays an important role in assessing various proposals for the interpretation of quantum mechanics; see Albert (2023).

The space-time arena in which fundamental ontology is located and fundamental properties instantiated is also an element of the package. Like the other parts of the package, it earns its place in the best system by its role accounting for the macroscopic phenomena with which physics begins and whatever else is introduced as elements of that account. The space-time with which physics begins consists of three spatial and one temporal dimension of our ordinary experience. But neither the dimensionality nor the geometry of the fundamental arena is metaphysically given a priori. It might be that there is no fundamental space-time arena, but spatial temporal relations are fundamental. Or it might even turn out that neither space nor time nor the geometry of space-time are fundamental but that they emerge from the ontology, properties, and laws in the best package.

Summarizing: the PDA, like the BSA, is based on the idea that there exists an optimal systematization of the distribution of fundamental ontology and properties, and that this systematization determines the fundamental laws. But unlike the BSA, the PDA does not assume the existence of Lewis's perfectly natural properties and a Humean mosaic at the outset. Instead, it involves a system of fundamental properties, entities, laws, and space-time all together as described by the world's optimal theory of everything. The PDA does this by explicitly building into the criteria for a candidate TOE that it provides the basis for the systematization of macroscopic facts described in the languages with which science begins. Especially important are the motions of ordinary physical objects described in ordinary languages. These languages are ultimately rooted in human biology and psychology. It is this requirement that replaces the role that the language of perfectly natural properties plays in the BSA. According to the PDA, laws are consequences of an optimal systematization of facts that anchor and are anchored in the macroscopic phenomena with which physics begins. Probabilities are discussed in the next chapter and special science laws in the chapter after that.

Because the fundamental ontology and properties of the world's best system need not be metaphysically fundamental in the way Lewis's perfectly natural properties are supposed to be, I will call them "scientifically fundamental." Because the distribution of the scientifically fundamental ontology and properties plays the role that the Humean mosaic plays in Lewis's BSA by providing a supervenience base for all truths, I call it "a scientific mosaic" (SM). It may not be unique since there may not be a unique collection of scientifically fundamental properties. Unlike Esfeld's super-Humean mosaic, the SM includes the instantiation of scientifically fundamental entities and properties which might not be propertyless enduring particles. It may, or may not, include particle trajectories depending on whether the PDA best system includes particle trajectories in its fundamental ontology. But whatever fundamental ontology and properties it includes, it will need to account for ordinary macroscopic objects and their motions and something like particles will certainly figure in this account.

The condition that a best system's fundamental ontology provides the basis for explaining the motions of macroscopic objects immediately excludes ExFx from being the axiom determining a best system. While ExFx is maximally informative in the sense that it excludes all possible worlds except the actual one, it doesn't provide information in way that enables the extraction of information about the motions of macroscopic objects expressed in ordinary scientific languages. It is of no use in predicting or explaining the motions of ordinary macroscopic objects or anything else of interest to science. The fact that a best system provides accounts of the motions of macroscopic phenomena as described in ordinary scientific languages also penalizes gruesome predicates and keeps the best systems' languages from being gruesome as long as macroscopic phenomena are characterized in ordinary non-grue terms.

There are some important further differences between the PDA and the BSA. One is that the PDA doesn't require that scientifically fundamental properties are categorical and so it allows that candidate systems posit scientifically fundamental properties with necessary connections, fields, fiber bundles, and other exotica. Scientifically fundamental ontology is determined by what makes for the best systematization as opposed to a priori metaphysics.

Consequently, on the PDA there may be truths stating necessary connections between scientifically fundamental properties. But these necessities shouldn't disturb sensible Humeans, since the necessity is imposed from above by the best systematization. This contrasts with the powers view, according to which a law is a regularity resulting from the operations of powers. My version of a Humean account of laws is not committed to there being no necessary connections between properties, only to laws not being determined by necessary connections between scientifically fundamental properties.[13]

The metaphysics that goes along with the PDA is a version of metaphysical monism. It declares that reality as a whole is metaphysically fundamental, and that it can be structured in many different ways. On this account, metaphysical fundamentality is not what those who have a "building block" conception of reality have in mind. On the building bock account, there is a unique collection of fundamental entities (e.g. individuals, properties, relations, etc.), and a unique way of putting them together that makes the world. Lewis is one of those who has such an account. He hopes that metaphysical and scientific fundamentality coincide, but we have seen that scientifically fundamental properties may not match his metaphysical building blocks—perfectly natural properties—and that there may be more than one collection of scientifically fundamental properties. The PDA doesn't reject or accept this kind of metaphysics, but it doesn't need it. Maybe Lewis's perfectly natural properties really exist, but we can wonder with van Fraassen why a scientifically optimal fundamental theory's primitive predicates should refer to them and, with Dasgupta, why they are more valuable than other properties whose instantiations serve as a supervenience base for non-fundamental facts. On this alternative to the building block view reality is one whole that can be described in many different ways with different fundamental scientific structures.

This is a good place to enumerate the differences between Lewis's perfectly natural properties and the PDA's scientifically fundamental

---

[13] Heather Demarest has proposed a different way of combing a best systems account of laws with necessary connections. According to Demarest, fundamental properties are powers or potencies that necessarily connect objects that instantiate them. Laws are axioms of the system that best systematizes the distribution of properties in all worlds in which the fundamental powers are instantiated. This is an ingenious idea that enables powers accounts to avoid the mismatch problem and also solves the problem of combining multiple powers. But unlike the PDA, it is genuinely anti-Humean since part of what makes a generalization a law is that its predicates refer to properties that are necessarily connected. In contrast, the PDA counts as Humean since what makes a generalization lawful is its being entailed by an optimal systematization. If the properties referred to by predicates in a generalization are dispositional, it is because they appear in the optimal systematization.

properties and the differences between the BSA and the PDA. Recall the listing of the features of Lewis's perfectly natural properties:

(i) They are a unique small subset of all properties, i.e. are sparse.
(ii) Which properties are perfectly natural is an objective matter of metaphysics independent of human interests.
(iii) The distribution of perfectly natural property/magnitude instantiations throughout a world's space-time is a supervenience base for all truths at that world.
(iv) They are categorical and intrinsic to the points or individuals at which they are instantiated.
(v) They are instantiated at space-time points or to concrete point-sized individuals.
(vi) They are the basis of an account of degree of naturalness for all properties and an account of objective similarity.
(vii) Every possible world is a possible space-time arena decorated with instantiations of perfectly natural properties and every space-time arena and every such a decoration is a possible world.
(viii) Not every perfectly natural property is instantiated at the actual world. The properties not instantiated at a world w he calls "alien" to that world.
(ix) They are reference magnets that play a central role in accounts of the reference of mental and linguistic representations. Degrees of naturalness play a role in determining reference by selecting references of predicates from among candidates that satisfy other criteria in terms of their naturalness.
(x) One of the aims of fundamental physics is the discovery of perfectly natural properties. Possible examples of perfectly natural properties are mass, charge, and spin.
(xi) Thay are the referents of predicates in languages in which candidates for a world's scientifically optimal systematization are formulated. Laws are theorems of the true systematization and so refer to perfectly natural properties.
(xii) They are the basis for solving Goodman's "New Riddle of Confirmation," since perfectly natural properties are projectible and the more natural a property, the more projectible it is.

This is what the PDA says about scientifically fundamental properties:

(i) They are sparse but may not be unique since there may be alternative sets of scientifically fundamental properties. There may be more than one set of scientifically fundamental properties corresponding to different best systematizations.
(ii) Which properties are scientifically fundamental is an objective matter determined by optimal systematizations although criteria for determining an optimal systematization have been devised within science.
(iii) The distribution of scientifically fundamental property/magnitude instantiations throughout a world's space-time is a supervenience base for all truths at that world.
(iv) Scientifically fundamental properties need not be instantiated at space-time points but can be instantiated at larger regions, even the entire universe.
(v) They need not be categorical or intrinsic to points or individuals at which they are instantiated. Properties instantiated in different regions may be necessarily connected.
(vi) They need not but may provide an account of similarity and degrees of naturalness that matches intuitions about these notions.
(vii) Since they need not be categorical and instantiated at points, they need not satisfy the recombination principle and provide the basis for an account of possible worlds. The PDA is not committed to any account of possible worlds or metaphysical possibility or the existence of alien properties.
(viii) Like perfectly natural properties, scientifically fundamental properties are the referents of the predicates of languages in which candidates for a world's scientifically optimal systematization are formulated. But the PDA allows for laws that are formulated in terms of properties that are not scientifically fundamental.
(ix) They can but need not play the role of reference magnets in a theory of reference.[14]

---

[14] The PDA isn't committed to any theory of reference, but it does suggest that reference facts about the macroscopic terms with which we start are prior to and play a central role in determining the references of more fundamental terms.

While the PDA is Humean in spirit it doesn't satisfy the letter of Humeanism as construed by Lewis, since it allows that there may be scientifically fundamental properties/quantities instantiated in distinct regions that are necessarily connected. It permits this if an optimal scientific systematization of first-order truths involves fundamental properties/quantities that possess necessary connections. But the PDA satisfies the Humean spirit, since according to it, laws do not govern or necessitate regularities but rather they are laws because they are part of an optimal systematization. The PDA grants that physics and other sciences are committed to nomological necessity and possibility, counterfactuals, and objective probabilities but these are not fundamental. Rather, as on Lewis' BSA they supervene on the distribution of fundamental properties—the scientific mosaic. Also, their semantics may be specified in terms of possible worlds, but the PDA has no need for metaphysically fundamental possible worlds as truth makers for them. While the PDA allows for necessary connections they arise, if they do, because they aid in unification. Humeans who think that laws explain by systematizing and unifying should have no objections to them.

Another important difference between the BSA and the PDA is that, in the PDA, the language in which candidates for best system are formulated include not only expressions definable in terms of fundamental expressions, but also terms for macroscopic properties that cannot be easily defined in terms of scientifically fundamental predicates and vague terms if these satisfy the criteria for an optimal system.

Special science laws involve terms referring to properties that are not scientifically fundamental. Such terms are also involved in specifying probability distributions over fundamental states. Also, the best system for our world plausibly includes a description of the macroscopic state of the universe at an early time usually called the Past Hypothesis" (PH) that implies that the entropy of the very early universe was extremely low.[15] According to the PDA, the language in which a best system of the

---

[15] The PH plays a very important role in accounting for the arrows of time, thermodynamics, and other macroscopic phenomena. Unlike the fundamental dynamical laws, it is stated in macroscopic language and is vague. Laws are propositions entailed by the best system of our world and so are not, as in Lewis's account, restricted to just generalizations. But we may want to restrict laws to those propositions that do or can figure in scientific explanations.

world is formulated includes terms referring to macroscopic properties and relations.

Let's return to some other considerations concerning the PDA of laws. We might worry that it is risky to build an account of laws that depends on the existence of a TOE or best system since the actual world may not have one. I think that the worry is premature since there is good reason to think that the world does have a TOE or a systematization close enough to sustain the PDA account of laws. Physicists have not yet found a TOE since there is, as of now, no theory that satisfactorily combines general relativity with quantum mechanics and that applies at very high energy scales and for very strong gravitational fields and covers the very early universe, black holes, and dark matter. What the fundamental ontology and laws of such a theory turn out to be and even if there is such a TOE is not known. But it is plausible that the standard model of effective quantum field theory together with general relativity—what Frank Wilczek calls "core theory" implies all the lawful macroscopic regularities that hold outside of high energy and strong gravitation. Sean Carroll writes:

> I will argue that we have good reason to believe that this model is both accurate and complete within the everyday-life regime; in other words, that the laws of physics underlying everyday life are, at one level of description, completely known. This is not to claim that physics is nearly finished and that we are close to obtaining a Theory of Everything, but just that one particular level in one limited regime is now understood.[16]

By "everyday life," Wilczek and Carroll mean not just everyday macroscopic phenomena but also phenomena and laws of special sciences. Their reason for believing this is that the core theory explicitly specifies

---

[16] Carroll (2022). In addition to the quantum field theory of the standard model and general relativity, a theory capable of accounting for special science laws and macroscopic phenomena needs to include laws that underlie statistical mechanics. One way of doing this has been developed by David Albert and myself that consists of adding a probability distribution over the dynamically possible histories of the universe. We call this package "The Mentaculus"; see Albert (2000); Loewer (2021) and (forthcoming b). It is likely that any adequate best system will include the Mentaculus.

the physical situations in which it holds, and the special sciences apply within these limits. This supports the existence of a TOE of our world, or a theory close enough to play the role of a TOE determining the fundamental laws. While the fundamental ontology and laws of a true TOE may differ from current proposals for fundamental ontology and laws, it is very plausible that current proposals can be absorbed into a final theory.

It is a presupposition of the success of physics, at least since Newton, that our world does have a systematization along the lines of a TOE. Nothing guarantees this, neither theology nor metaphysics. Science is a risky enterprise. So far, taking the risk has paid off.[17] But we don't want to say that if it turns out that our world fails to have a TOE, there are no laws. So, I suggest that the PDA is modified to say that even if the world doesn't have a theory that covers all physical phenomena, the laws are determined by a theory that comes closest to being a TOE as long as it is sufficiently close.

We can illuminatingly contrast the PDA with Lewis's BSA by adopting Lewis's account of properties.[18] On this account, every class of possible individuals, including space-time points, is a property. A select number of these are perfectly natural properties and some of those are instantiated in the actual world. Lewis holds that these properties are instantiated at space-time points (or individuals no bigger than a space-time point) and are categorical. The space-time distribution of their instantiations forms the Humean mosaic. Recall that the Humean mosaic is a supervenience base for all actual truths including laws, counterfactuals, chance propositions, and so on. Think of the perfectly natural properties as glowing with a kind of metaphysical light. Lewis thinks that it is the goal of physics to find them and the best systematization of their instantiations.

In contrast, on the PDA, there may be many sets of properties that can serve as supervenience bases for all truths. Physics aims to find an optimal systematization of the macroscopic truths of interest to

---

[17] But of course scientific developments, especially technological developments, have led to other risks. These are primarily due to there not having been a parallel development in the humanities and ethics.

[18] This way of comparing the two accounts was suggested in conversation with Alyssa Ney.

humans, and especially human scientists—for example, the truths of thermodynamics. This involves additional non-fundamental ontology and properties—for example, molecules and their properties. A set of scientifically fundamental properties is a minimal supervenience basis for actual truths in an optimal systematization. These properties glow not with a metaphysical light but with a kind of scientific light. Unlike the set of supposed perfectly natural properties, there may not be a unique set of scientifically fundamental properties but more than one set which is optimal.

The set of perfectly natural properties, if there are such, might be identical to a set of scientifically fundamental properties or it might not be. In fact, no set of scientifically fundamental properties might be identical to the set of perfectly natural properties. This is the mismatch problem pointed out by van Fraassen and Demarest. It is easy to see why it is valuable for scientists to find a set of scientifically fundamental properties, so there is not a missing value problem as there is for perfectly natural properties. In fact, unless it can be shown that metaphysical perfectly natural properties serve some explanatory purpose, their light is extinguished by scientifically fundamental properties and we may doubt that they are needed or exist.

A way of understanding the PDA is that it takes the idea from Lewis's BSA that propositions are laws in virtue of their place in a system Σ that unifies and systematizes, and takes the idea from Quine that science "starts in the middle" with the macroscopic facts about the world of interest to science. The middle facts get the systematization going. Thinking about laws this way is completely consonant with Quine's epistemology of science but takes a metaphysical turn. The PDA replaces the metaphysics on which Lewis's BSA is built and replaces it with a metaphysics inspired by Quine's account of epistemology. Lewis was Quine's student. He thought that the Quinean project of explaining the world must be like in order for science to be successful requires a metaphysics of natural properties and an account of metaphysical modality involving possible worlds. The PDA shows that no such metaphysical additions are required by science if it is understood along Quinian lines.

8
The PDA and Chance

Probability is the most important concept in modern science, especially as nobody has the slightest notion what it means.[1]

In this chapter, I show how objective probability can be incorporated into the PDA. My account derives from Lewis's account of chance but there are important differences and improvements. First, a little background.

Data accumulated from census, agriculture, insurance, etc. during the seventeenth and eighteenth century made it apparent that many processes shared with gambling the feature that even though it is not possible to predict what will happen on any occasion, it is possible to accurately predict approximately what will happen "in the long run." Examples include the relative numbers of male and female births each year, the average yield of crops, average rainfalls, suicides per year, and so on. In the case of throwing dice, it is possible to assign a number—the probability—to an individual toss and estimate accurately the frequency that would result from many tosses. Similarly, one can assign a probability to the next child born being male and accurately estimate the approximate frequency of male births during a year. The mathematics needed for performing such calculations initiated by Pascal was developed by Bernoulli and later by Laplace, Gauss, and Poisson who described various probability distributions and formulated the laws of large numbers.[2] These mathematical laws describe how, as the number

---

[1] Bertrand Russell (1929); cited by Alan Hájek (2019). This article is an excellent introduction to various interpretations of probability.
[2] The history of the development of the mathematics of probability goes back prior to Pascal to the Middle Ages with figures including Maimonides and Roger Bacon, and before that to the ancient world.

of repetitions of a type of trial increases, the frequency of a type of outcome stabilizes.

The weak law of large numbers says that, for sufficiently large sample sizes, assuming that the trials are independent, there is a very high probability that the average of sample selected at random from a population will be close to that of the population mean. The difference between the two will tend towards zero, that is, the probability of getting a positive number ε when we subtract the sample mean from the population mean is close to zero when the size of the observation is large. The weak law is used, for example, to calculate the probability that the frequency of heads among many tosses of a fair coin will diverge by a given amount from 0.5. Notice that the law doesn't say that the frequency strictly converges to the mean as the number of trials increases but describes how the frequency stabilizes in terms of probability. This is an important point to which we will return.

It is often said that non-trivial objective chance and determinism are incompatible. The dynamical laws of classical mechanics (and some versions of quantum mechanics) are deterministic. This raises puzzles about how probability can describe actual events if these theories are correct. Laplace explained determinism in terms of what a superintelligence could predict. He says that if determinism obtains then a superintelligence who knows the state of an isolated system (or the entire universe) at one time could calculate what it will be later times.[3] Removing reference to knowledge, determinism says that the laws and the state of any system, including the entire universe, at a time t imply the state of the system at any other time t*, as long as the system remains isolated.

If the superintelligence knows only the macroscopic state of the system, the fundamental laws won't enable him to calculate its later macro states. However, if he knows a probability distribution over the system's microstates compatible with its macroscopic state, he can calculate the

---

[3] "...given for one instant an intelligence which could comprehend all the forces by which nature is animated and the respective situation of the beings who compose it—an intelligence sufficiently vast to submit these data to analysis—it would embrace in the same formula the movements of the greatest bodies of the universe and those of the lightest atom; for it, nothing would be uncertain and the future, and the past, would be present to its eyes"; Laplace (0000) 4.

probability that it will evolve in accordance with the dynamical laws into various future microstates and macro states. If he knows which macro states different microstates realize, he will be able to calculate probabilities for future macro states given the earlier macro state. Since the fundamental dynamics are assumed to be deterministic this suggests that these probabilities must be epistemic.

Laplace seemed to have understood probabilities as subjective degrees of belief reflecting a particular person's actual knowledge. But in certain situations, degrees of belief also seem to be objective, reflecting the degrees of belief that a person objectively ought to have in that situation. This is how Boltzmann seems to have understood probability in statistical mechanics. These epistemic probabilities are not merely subjective, reflecting ignorance of a system's microstate, but apparently arise from the objective features of the particles that constitute the system. Boltzmann and Maxwell attempted to derive them from assumptions about how gas particles interact, given classical mechanical laws. But their derivations employed probabilistic assumptions. Exactly what statistical mechanical probabilities are remains a vexing issue.

Ian Hacking observes that probability is Janus-faced. One face looks to the world reporting an objective feature, and the other to the mind recommending the degree of ignorance or belief. That degree of belief is the degree of belief one rationally should have given the objective probability. According to Hacking:

> Poisson and Cournot used the words *chance* and *probabilité* to name two concepts. Probability would mean credibility, or degree of reasonable belief.... But chance will denote an objective property of an event, the 'facility' with which it can occur.[4]

Cournot realized that there must be a connection between chance and probability. In order for chance to apply to the world and enable the inference of chances from events, there must be a principle connecting statements about chance with statements about ordinary events. He expresses the connection this way:

---

[4] Hacking (1990) 96.

A physically impossible event is one whose chance is infinitely small. This remark alone gives substance—an objective and phenomenological value—to the mathematical theory of probability.[5]

The idea behind this remark is formulated as what has been called "Cournot's Principle." One version of it is:

(CP) if an event E has very small probability it will not happen.[6]

There are obvious problems with this version of CP. Since there are situations in which every event that can happen has very small probability, CP results in the lottery paradox that none will happen when one must happen.

There are formulations of CP that add further conditions in addition to low probability that avoid this paradox but I won't discuss them further since David Lewis formulated an account of chance that connects chances to the world, and formulated more sophisticated principles which connect chances with degrees of belief about the world that also avoid the lottery paradox. We will examine his account of how chance connects with the world shortly, but first we will look at principles connecting chances with degrees of belief. The most important is the "Principal Principle":

$$(PP)\ Cred(E/A\&(P(E) = x)) = x$$

PP says that a rational person's conditional degree of belief for E, given A and that the chance of E is x is x. A is information that is "admissible" with respect to E. An initial gloss on "admissible" is that A is admissible with respect to E iff A provides information about E only by way of providing information about P(E). An example of inadmissible information is information provided by a crystal ball announcing that E is true. I will examine some other formulations and reformulations of the Principal Principle later that dispense with admissibility.

---

[5] Cournot (1843).
[6] See Shafer (2006) for an extensive discussion of Cournot's Principle and its history.

Lewis says that what we know about chance is that it is the property or function of propositions, if there is one, that satisfies the PP or a very similar principle that connects chance with credence. So, the question is what kind of property can an event E possess that both is objective and also recommends the rational degree of belief that E occurs? This property is not a usual physical property like mass or charge in that, on the one hand, it is an objective feature of physical events, and on the other, has a normative epistemic dimension.

Some thinkers who have considered this question think that no objective physical property can be like this. Bruno de Finetti was one. He tried to account for the appearance of such a property by showing how in simple cases, if a person's subjective degrees of belief about E satisfy certain conditions, for example exchangeability, it is as if the person believes that there is an objective chance of E even when in reality there are no objective chances. If a person's degrees of belief are exchangeable, there is no need to appeal to the existence of objective chances to account for how she ought to form and change her degrees of belief when obtaining evidence.

The problem with this approach as an account of the chance statements that occur in scientific theories, like quantum mechanics and statistical mechanics, is that these chances are independent of what anyone believes and are involved in causal explanations of physical events. It isn't plausible that the chance of a radium atom emitting an alpha particle during a certain time period is just someone's subjective degree of belief no matter how exchangeable her degrees of belief may be.

Another approach that puts chance on the mind side is objective epistemic accounts that say that logic or rationality dictates what degrees of belief one should have in certain situations. This seems to make probability objective in the same way that principles of logic are objective and account for why a person's degrees of belief should conform to these objective probabilities. This idea goes back to the early days of probability which invoked the principle of sufficient reason or the principle of indifference to determine probabilities.

The principle of indifference says that if one knows that there are a certain number of possibilities but has no further information, it is rational to assign each possibility equal probability. A related non-epistemic principle is that symmetric possibilities have equal probability. These principles have been developed by Carnap, Jaynes, Williamson, and others.[7] There are two main problems with this type of approach to the chances that occur in physical theories. One is that there is typically no uniquely correct objective way of carving up a space into possibilities.[8] More worrisome, even if there is a unique carving, it is not clear why that should justify assigning equal epistemic probabilities to each possibility. Perhaps the rational state of mind is one better described by "I don't have a clue." Why should symmetric possibilities have the same probabilities rather than no probabilities at all?

I have called objective probability "chance," but it may be better to call it "physical probability." The reason is that the term "chance" is often limited to physical probabilities that are incompatible with determinism and are specified by indeterministic laws. Lewis himself thought of it this way. But, as we will see, there may be objective physical probabilities even when the fundamental laws are deterministic.[9] So, I will continue to use the term "chance" for objective physical probability.

Among those philosophers who think that chance is an objective feature of the world, there have been several proposals about what it might be. As in the case of laws, these proposals divide between Humean and non-Humean accounts. On Humean accounts, chances supervene on the distribution of actual events while on non-Humean accounts, such chances are properties or magnitudes of a property, or are given by laws that are over and above the distribution of actual events. Some non-Humean accounts connect chances with governing laws and others consider them more like powers or degrees of powers.

One issue concerning chances is whether they apply to token events (events occurring at specific times) or to event types or to both. When we say that smoking increases the chance of heart disease, we are applying chance to types. But in "the chance as of today that Jones will win

---

[7] Carnap (1945); Jaynes (1968); Williamson (2010).   [8] See van Fraassen (1989).
[9] Schaffer (2007) argues that there are no non-trivial chances if determinism is true since the concept chance as he understands it has commitments that are incompatible with determinism. Whether to call the objective probabilities that occur in physics with deterministic dynamics chances or not seems to be merely a semantic issue. As we will see, the BSA account of chance applies uniformly to both deterministic and indeterministic objective probabilities.

next year's election is 0.7" we are talking about the chance of a particular event at a particular time. Some accounts hold that chance applies first to types, and if it applies to tokens at all it applies only relative to a type. Other accounts hold that token events have chances and chances apply to types derivatively. Some of the metaphysical views of chance I will survey apply to both.

The simplest Humean account of objective probability is the actual frequency account. It identifies the chance of event type E with the actual frequency of occurrences of events of type E on some kind of trial. As an account of the probability of the type of event "lands heads when flipped," it specifies the actual frequency of the coin landing heads when flipped. The obvious problem applying this to token events is that a token event belongs to many different types with different frequencies. The frequencies of a quarter landing heads on the next flip may differ depending on which type the next flip is included in. But there is no end of types with different frequencies in which it can be included.

Also, chances just don't seem to be the same as actual frequencies. The chance of a quarter landing heads may be 0.5 but the frequency of heads on flips of the coin might diverge from 0.5. The way we understand chance, the chance of the coin landing heads and the frequency with which it lands heads may always differ and even diverge considerably. This is the primary reason against the view that chances are actual frequencies but there are many others.[10]

Hypothetical frequency accounts identify the chance of a trial of type T resulting in outcome of type O with the limiting frequency of O that would result from unlimited many repetitions of T. These accounts can either be Humean or anti-Humean, depending on whether the hypothetical conditional has Humean truth conditions. Hypothetical frequency accounts permit actual frequencies and chances to diverge but face other problems.[11] I will not discuss frequency accounts further and

[10] Hájek (1997) provides fifteen arguments against actual frequentism.
[11] One problem is that it is not clear that the hypothetical "if the coin were flipped infinitely many times, the frequency of heads would be p" is evaluable. Also, without an account of how the relevant hypothetical circumstances are determined, the account is incomplete and not clearly Humean.

refer the reader to Alan Hájek's excellent papers that discuss the problems with both actual and hypothetical frequency accounts.[12]

On the clearly non-Humean side of the debate is the usual version of the propensity account, associated with Karl Popper,[13] according to which chance is a measure of the degree of the propensity of a process to produce a certain outcome. Propensities are like dispositions or powers, and chance is the degree to which that disposition operates. It is understood as metaphysically fundamental. On propensity accounts, chances don't supervene on the Humean mosaic but are elements of reality over and above it. This kind of account seems to be Pierce's view and was explicitly proposed by Popper as an account of quantum mechanical chances. Propensity chances are like non-Humean deterministic laws and powers in that they are involved in bringing about subsequent events. But unlike them, they are not "sure-fire" but rather select among alternative possible futures in accordance with the values of their chances. Propensity accounts don't try to characterize chances in terms of frequencies or degrees of belief but include propensities among the fundamental furniture of the universe and attempt to explain how they are connected to frequencies.

Since on propensity accounts there are no necessary connections between chances and frequencies, they allow the propensity chance of an event and the frequency of its occurrence to diverge arbitrarily far apart unlike actual frequency accounts. This, at first, is a point in its favor but it leads to problems. Before discussing them, I want to discuss a less serious objection to propensity accounts. The objection is the so called "Humphrey's paradox" which concerns the fact that degrees of propensity don't conform to the axioms of probability. The problem is that degrees of propensity violate the principle that if the conditional probability $P(B/A)$ is defined so is $P(A/B)$. While there may be a propensity chance of a radium atom emitting an alpha particle at a time, there may be no propensity chance of the alpha particle having been emitted previously by a radium atom. But according to Kolmogorov's axioms whenever $P(B/A)$ is defined and positive, so is $P(A/B)$.

---

[12] Hájek (1996) and (2009).   [13] Popper (1959).

Humphrey's paradox has been taken by some as sufficient reason to reject propensity theories of chance. That is not my reason. It is not difficult for advocates of propensity accounts to deal with the alleged paradox by modifying the probability axioms or restricting the application of the propensity theory by holding that not all objective probabilities are propensities. I won't discuss these proposals further since there are more serious metaphysical issues and problems with propensity accounts.

One problem is an analogue to van Fraassen's inference problem for Armstrong's account of laws. Recall that van Fraassen pointed out that the existence of Armstrong's relation of contingent necessitation between properties F and G does not entail that all Fs are Gs. This must be posited as a fundamental necessary connection. Similarly, there being a propensity of degree x for an F to result in the outcome of a G does not entail any value for the actual frequency of Gs among Fs. How do propensities result in stable frequencies? A related problem is that if degrees of propensity are objective probabilities, they should guide rational credence. This connection is formulated by a version of Cournot's Principle or the Principal Principle. But it is difficult to see why a relation or property of events that doesn't supervene on the Humean mosaic should guide rational credences about events in the mosaic. Lewis puts the point this way:

> Be my guest—posit all the primitive unHumean whatnots you like. (I only ask that your alleged truths supervene on being.) But play fair in naming your whatnots. Don't call any alleged feature of reality "chance" unless you have already shown you have something, knowledge of which could constrain rational credence. I think I see dimly but well enough, how knowledge of frequencies and symmetries and best systems could constrain rational credence. I don't begin to see, for instance, how knowledge that two universals stand in a special relation N* could constrain rational credence about the future co-instantiation of those universals.[14]

---

[14] Lewis (1984) 484.

So, the main problem with the propensity account is that, by making chances non-supervenient on the actual events, it is difficult to understand why principles that connect chances to actual events and beliefs about actual events like Cournot's Principle and the Principal Principle should hold.[15]

The difficulty of providing an adequate account of chance could tempt one to try to formulate scientific theories without it.[16] But its use is so widespread in science and everyday life that it is difficult to see how this could work. Fortunately, Lewis proposed an account of chance, a development of the BSA of laws, which seems to account for how chances can differ from but still be related to actual frequencies and apply to token events. At the same time, he hopes the account can explain how chance constrains rational credence. Furthermore, as we will see, contrary to his own view, this account allows for objective probabilities that are compatible with determinism. Lewis's proposal is that chances come together with probabilistic laws that specify chances in a theory that optimally systematizes the Humean mosaic. As a consequence, chances supervene on the HM. The idea is that since chances supervene on the HM, they will be connected to actual frequencies and that this will enable them to guide rational credences about them. Lewis's account works only for objective probabilities that are specified by laws but, since as I will argue, there is reason to believe that all objective physical probabilities are like this, it is not a limitation. I will argue that while Lewis's account requires some modification and development, it provides the basis for an account of objective probability that is superior to any of the accounts that we have just canvassed.[17]

Here is how Lewis adds probability to the BSA. A function term, "$P(E,t)$," denoting a function from propositions and times to real numbers between 0 and 1 is added to the language for candidates for best system to enhance informativeness while minimizing added complexity.

---

[15] For more problems with propensity accounts of chance, see Eagle (2004).

[16] An intriguing idea is to replace probability with the concept of typicality and make sense of it. This program has been developed by Dürr, Goldstein, Maudlin, Wilhelm, and others. I discuss and critique it in Loewer (forthcoming b).

[17] A similar Humean account of objective probability is developed by Carl Hoefer (2019). Unlike Lewis, Hoefer separates it from a Humean account of laws (which he rejects) and applies it to subsystems rather than to the entire universe.

Truth conditions for sentences containing the expression are then provided by specifying criteria for evaluating systems containing these sentences. The idea is that P(E,t) is true iff the best system of the world entails it. Here is an illustration of how this works. Consider a long sequence of the outcomes of coin tosses, for example, <HHTHTTTHHTTHTHHTTT...>. A very informative description of the sequence specifying each outcome would be very complicated. A simple description, for example, "approximately half Hs and half Ts" would be much less informative. But the description "outcomes of independent trials with probabilities P(H) = 0.5, P(T) = 0.5" is fairly simple and much more informative. If it is the best system for the sequence then P(H/T) = 0.5.

Understood this way, chances are not fundamental features of events. Rather, a function expression specifying chances is introduced into theories that are candidates for optimal scientific systems to provide information about the mosaic. The proposition "the chance of E at t is x" says that the actual mosaic is such that its scientifically optimal systematization(s) entail(s) that the chance of E at t is x. In this way, chances are given by BSA laws that specify chances.

Lewis applies this idea only to chances that are specified by a certain kind of indeterministic dynamical law. These are laws of the form "if h is the actual history of the universe or an isolated subsystem up until and including time t, the chance of s at t is x." In all the examples of fundamental theories I know with indeterministic dynamical laws, the dynamical chances at t are determined by the state just at t. The history prior to t is irrelevant. Sequences that satisfy such laws are called "Markovian."[18] The history prior to t is irrelevant so I will consider laws of the forms $P_t(s(t')/s(t)) = x$ where s(t) is the state at t and s(t') the state at a subsequent time. The law may also take the form of a differential equation that describes the evolution of chances. Lewis's proposal is to permit axioms of candidates for best system Σ to include dynamical laws of this form. For such theories, the possible histories of the universe exhibit a branching structure in which at each branch point the laws

---

[18] Lewis's account allows for non-Markovian laws and there have been proposals for fundamental laws that are non-Markovian, but I will ignore them.

determine the chances of futures of that branch.[19] The values of chances at a world are specified by the dynamical laws of the best system of the world. For example, the chance now of it raining tomorrow is the sum of the chances of each of the branches emanating from the macro state of the world now on which it rains tomorrow.

There are two related issues that must be addressed by this account. First, we need to describe how assigning chances provides information about the Humean mosaic. Second, we need to describe how candidates for the best system are to be evaluated to balance informativeness, simplicity, and whatever other criteria are deemed scientifically important to determine optimal systems. Although Lewis never explicitly makes this connection, it is plausible to think that chances provide information about the mosaic by way of the Principal Principle. A chance theory does not say what will happen, it rather specifies what credences one who has no inadmissible information should have about what will happen. A weatherman who tells you that the chance of rain is 0.8 is telling you that you should have 0.8 credence in its raining and thus provides information about the weather. This makes satisfying the Principal Principle constitutive of Lewis's account of chance. It supports his claim that what we know about chance, whatever it is, is that it is connected with credences by the Principal Principle or something close to it.

To evaluate candidate systems with probabilistic laws, Lewis proposes that they aim to maximize what he calls "fit." The higher the chance a system assigns to a mosaic the better it fits it. So understood, fit is a kind of informativeness appropriate for chance. The better a theory fits the facts the more it says about those facts.

Lewis's proposal of fit as a criterion for evaluating probabilistic systems seems plausible since it reflects the "likelihood principle" (LP) that is central to Bayesian and other approaches to statistical inference. The LP says that given alternative equally credible exclusive theories $\Sigma 1$, $\Sigma 2$, $\Sigma 3$,... and evidence E, other things being equal, one should have

---

[19] An example of such a law is the dynamical law for the evolution of quantum states proposed in the GRW theory. See Albert (1992) for discussion.

greatest confidence in the Σ on which E is most likely.[20] If E is the Humean mosaic, and the Σs are alternative candidates for best system, then the principle says one should have greatest confidence in the system Σ that best fits the mosaic. It seems reasonable that this is the system that provides the best information about the mosaic.

But there is a problem on the horizon. Lewis noticed that there is an apparent conflict between his Humean BSA and the Principal Principle. He calls the conflict "the big bad bug."[21] The problem is that the best system Σ for a mosaic M may assign a positive probability to a proposition that describes another mosaic M*, even though M* undermines Σ so that the truth of M* falsifies Σ. For example, if M consists entirely of a trillion-member sequence of randomly distributed Hs and Ts with half of each, then the best system Σ for M assigns P(h) = 0.5 and P(h/t) = P(h). Σ also assigns a positive, although exceedingly small, probability to a sequence of a trillion Hs. But a mosaic consisting of a trillion Hs excludes Σ since its best system is simply that only Hs occur. According to the Principal Principle, one's credence in a trillion Hs should be positive, but since a trillion Hs is excluded by Σ, it should be 0. An outcome of a trillion Hs *undermines* this system. The problem is that while the BSA allows for a great deal of frequency tolerance, since the chance theory supervenes on the mosaic, it also excludes mosaics to which its best system may assign a positive probability. The result is a contradiction between Humean supervenience and the Principal Principle that Lewis described as "a bug."

At first, the bug seems to be a big problem for Humean accounts of chance. But in fact, it is not much of a problem, and the bug can be easily squashed.[22] First, note that the problem arises only when the proposition describing the sequence of events explicitly specifies that it constitutes the entire mosaic. If it doesn't do that then the Principal Principle doesn't say that one's credence for it should be 0, since the proposition describing the sequence isn't inconsistent with the theory.

---

[20] On Bayesian accounts, if one assigns equal prior probabilities to exclusive and exhaustive alternatives h1, h2, then the posterior probabilities, given evidence E, depend only on the likelihoods P(E/h1), P(E/h2)....

[21] Lewis (1986) xiv.   [22] Hall (1994).

One response to the problem is to reformulate the Principal Principle to exclude propositions that say that they completely describe a mosaic. Alternatively, Hall, Thau, and Lewis suggest replacing the Principal Principle with the New Principle:

(NP) Cred(E/A&Σ&P(E,t) = x) = x

NP says that a rational person's conditional degree of belief for E, given that the chance of E at t is x and A and Σ is x. By conditionalizing on Σ, mosaics incompatible with Σ are excluded and so NP assigns them 0 credence. NP gives the same or almost the same results in almost all cases, as the Principal Principle does, and performs the job of providing information without conflicting with the BSA.

Although the big bad bug turns out not to be so big, the BSA still has the consequence that propositions that undermine a chance assignment may be given a positive chance. This is the probabilistic counterpart to the feature of the BSA that there are worlds that are compatible with a law that undermine its laws. These are inevitable consequences of Humean accounts of laws and probabilities. If it is a bullet, it must be bitten by any such account. Humeans can digest it.

It is often thought that if the world's dynamical laws are complete and deterministic then there is no room for objective physical probabilities. Those who think this, usually also hold that the probabilities that occur in scientific theories are merely epistemic. Lewis seemed to have thought this, since he held that his account of objective probability applies only to indeterministic laws. This view gained support from the fact that quantum theory, at least on widespread interpretations, includes probabilities connected to indeterministic laws. But, as mentioned earlier, there are theories in physics, like classical statistical mechanics and Bohmian quantum mechanics, that apparently include objective probabilities even though their dynamical laws are deterministic.[23] These probabilities don't seem to be merely epistemic.

A great advantage of the BSA is that it is possible to modify it to apply to objective probabilities, whether dynamical laws are deterministic or

---

[23] Loewer (2001) was first to notice this.

indeterministic. This is accomplished by adding to candidate best systems an axiom specifying a probability distribution or density over initial conditions compatible with the dynamics. This axiom qualifies as a component of a best system in virtue of its adding informativeness and fit at little cost in simplicity. The probabilistic laws of Bohmian mechanics and classical mechanics can be understood in just this way.

Since some of the ways we have come to think about chances are bound up with indeterminism, one may be prefer not to call deterministic objective probabilities "chances."[24] But that is merely a semantic matter. I will continue to refer to all objective physical probabilities as "chances." The important point is that they are objective probabilities, and they are compatible with deterministic dynamics. Further, they supervene on the mosaic and the conditional probabilities they entail guide credence in accord with the NP and related principles.

Best systems that include an axiom that specifies a probability distribution or density over all physically possible initial conditions and include laws whether deterministic or indeterministic have an interesting feature. These systems assign probabilities (or densities) over all physically possible histories of the universe. Because of this, they assign an objective conditional probability P(B/A) for every pair of propositions B and A whose truth conditions are classes of physically possible histories and for which P(A) > 0. This constitutes what I will call "a probability map of the universe."[25] Together with a conditional form of the PP or NP, the map determines objective conditional credences for all pairs A, B.

For a system Σ that provides a probability map of the universe, there is no need for the probability function to make explicit reference to time, and the conditional form of the Principal Principle or its replacement, the conditional NP (which I will shortly introduce) can be formulated without any reference to admissibility or to time. Lewis's indeterministic chances are the probability map chances conditional on the history of the universe up until and including t when the laws are

---

[24] Schaffer (2007).
[25] David Albert and I call such a probability map associated with statistical mechanics "The Mentaculus." Loewer (forthcoming b) describes it in detail and shows how it provides the basis for accounts of time's arrows, counterfactuals, objective degrees of credence.

indeterministic. Also, probability map chances are not temporally directed. There is an objective chance of B given A even when B may be temporally prior to A, and that chance may differ from 1 or 0 even when the events described by B have already occurred.

The PP and NP can be reformulated in terms of conditional probabilities:

(PP*) Cred(E/A) = P(E/A) where Cred(E/A) is the rational degree of belief that E occurs conditional on A and P(E/A) is the chance of E given A.

(NP*) Cred(E/A) = P(E/A&Σ) where Cred(E/A) is the rational degree of belief that E occurs conditional on A and P(E/A) is the chance of E given A and Σ is the true theory of chance.

Notice that PP* and NP* make no reference to admissibility. The reason is that conditional chances cannot be undermined by further information since that information is just added to the condition. When determining what degree of belief one should have in a proposition, one should conditionalize on all the information one has or should have. If the chance of snow tomorrow given the state of the world today is 0.73 but a crystal ball tells you that it will snow, then you should set your degree of belief in its snowing tomorrow to the conditional chance of snow tomorrow given both the state today and that the crystal ball says it will snow. If there is a probability map of the universe, it will assign an objective probability to its snowing tomorrow conditional on the history until now including that the crystal ball says it will snow tomorrow.

The BSA of chances applies in the first instance to states of the universe (or an isolated system) at a time and then to token events that are components of those states. On Lewis's account, the laws specify the chance at a time t of a particular event occurring at a later time t'. But on the extension of the BSA which includes a probability distribution over initial conditions, the laws yield a probability map. The map will imply the objective probability of a macroscopic event type A being followed within a certain time by an event of type B. Most of these conditional probabilities are unimportant and inaccessible, but in some cases

P(B/A & D) is the same for a wide variation in conditions D and we can know that probability. This, for example, is the case for the chance of an ordinary quarter landing heads when tossed. If the background macro state D is fixed as the toss being on earth, weather conditions not being severe, the planets being on their usual orbits, and no other large celestial objects in the solar system, etc. then this conditional probability is 0.5 pretty much whatever is going on elsewhere. In these cases, we can speak of the event type A resulting in event type B with a specific chance.

The BSA of objective probability seems to be what we have been searching for. It is an objective account of chance, but it doesn't suffer from the problems of frequency and propensity accounts. It applies to tokens and types, it is an account on which objective probabilities supervene on the mosaic, and thus it promises to explain why its chances constrain rational credence. If there is a probability map of the universe, it accounts for special science and other macroscopic objective probabilities. All this looks terrific. However, there are several problems with the BSA that need to be addressed.

The three most pressing problems are:

(i) Adam Elga pointed out that for mosaics with infinitely many events, like the actual world's mosaic, the chance of the mosaic will be 0 for every plausible candidate for best theory. This seems to render Lewis's account of it useless.[26]
(ii) Roman Frigg and others pointed out that it is possible to increase informativeness of a system by peaking its probability distribution around the actual mosaic without much sacrifice of simplicity. This raises the worry that Lewis's account won't yield a unique probability distribution.
(iii) Although Lewis may "see dimly" why the PP and NP should be satisfied, it hasn't been as clear to others. It would be desirable to have an argument that explains why BSA chances should satisfy the PP or NP. It turns out that the three problems are interconnected, and they can be mostly solved by the PDA's modifications of the BSA.

[26] Elga (2004).

The PDA incorporates Lewis's account of objective probabilities with the modifications I proposed, but it adds some further modifications which help deal with the problems faced by the BSA account. According to the PDA, candidate systematizations can contain axioms that specify probabilistic dynamical laws and laws that specify probabilities over initial conditions. Candidates for best system can employ macroscopic vocabulary in their formulation. As we will discuss in the next chapter, one of the criteria for evaluating candidates for best system is how well they systematize the special sciences as well as the fundamental ontology.

The criteria of fit can now be adjusted so that fit is measured in terms of the likelihood not only of the fundamental trajectory of the world but also of macroscopic systems of interest to science. This solves Elga's problem. Instead of looking at the exact microscopic history of the universe given candidates for best system, look at the likelihood of the totality of macroscopic systems of interest to science in a large region of space-time. These likelihoods will be small but greater than 0. The best system is the one that assigns the highest value to the totality of these macroscopic systems.[27]

Roman Frigg pointed out that the informativeness of a probability distribution can be increased by concentrating it around the actual world.[28] The cost in simplicity is small if the function specifying the probability distribution is simple. One problem with Frigg's suggestion is that while it may make the actual total frequency very likely, and alternative frequencies very unlikely, a too peaked distribution will assign probabilities to subsystems that are ill fitting. Related to this, a too peaked distribution gives very different results for counterfactuals. This is a worry since in the context of decision making it can imply that whatever action I take the actual result is likely. That is undesirable. The lawful probability distribution should be informative about counterfactual truths as well as actual truths. So, while more peaked distributions may be more informative about the fundamental history of the world, it

---

[27] This solution to Elga's problem is like the solution he suggested. He proposed that fit is measured in terms of certain "test" propositions. My suggestion is that these test propositions describe the macroscopic mosaic.

[28] Frigg (2008).

will not necessarily be more informative about macroscopic subsystems or counterfactuals. If it were, then it would be a better candidate for the best system.

While I know of no proof and suspect there cannot be one that following the PP or NP (or their conditional formulations) will always result in assigning high subjective credence to all and only true propositions, the PDA entails that a best system fits the world in the ways just discussed. Since satisfying the NP is constitutive of BSA and PDA chances, accepting Σ as a best system commits one to following the NP. Since the likelihood principle recommends Σ over its competitors, there is reason to accept Σ. The BSA and PDA allows one to see "dimly but well enough" why objective probability should guide credence via principles like the NP.[29] The conclusion of this chapter, so far, is that incorporating Lewis's account of objective probability into the PDA in the ways I suggested provides the best account of objective probability in town.

Metaphysical Coda

Whether the world's fundamental dynamical laws are indeterministic or deterministic has been thought not only to involve different understandings of probability, but also to correspond to deep metaphysical differences concerning the nature of time and existence. The metaphysical issue is the dispute between presentists and growing block accounts of time on the one hand, and eternalists on the other. Presentists hold that only facts about the present time exist, growing-block theorists add that past facts also exist, while eternalists hold that past, present, and future facts all have an equal claim to exist. Although a great deal of heat has been generated by this dispute, it is sometimes difficult to see what is really at issue. Eternalists employ an eternalist notion of existence

---

[29] Proponents of other accounts of chance can offer a similar argument for the PP. But the BSA has the advantage of explaining why the NP is constitutive of chance in virtue of its being the way chances provide information about the mosaic. An advocate of a propensity account can also claim that the PP is constitutive of propensity chances, but it is puzzling why this should be the case since propensity chances don't supervene on the mosaic.

whereas the other positions employ a notion of existence tied to time. Once this is made clear, the issue seems to come down to whether the eternalist notion of existence is legitimate.[30]

Lewis's BSA for both laws and chances assumes eternalism, since laws and chances are determined by the entire Humean mosaic which exists throughout all space and time. Future events play as much a role in determining the laws and probabilities as present and past events. From the perspective of the BSA and PDA, the differences between determinism and indeterminism and between indeterministic objective probabilities and determinist objective probabilities are not deep metaphysical differences. BSA and PDA laws and chances arise from a best systematization of the Humean mosaic. The mosaic is metaphysically fundamental. Neither the probabilities connected with indeterministic laws, nor those connected with deterministic laws are metaphysically fundamental. On the BSA and PDA, all physical probabilities are objective features of the world that have epistemic consequences by way of the PP.

---

[30] A problem with presentism and growing block accounts is that they assume that there is a fact about the present throughout all of space, and that seems to require that there is a fact about whether events are simultaneous. Theories of relativity posit a space-time which doesn't require such facts and that seems to throw a wrench in these metaphysical views. Those who want to maintain their views have responses to this, but I won't purse this aspect of the discussion since my main interest is how this metaphysical dispute interacts with accounts of objective probability.

9
Special Science Laws and the PDA

Most of my discussion so far has concerned laws and chances that occur in fundamental physics. There are also regularities that appear to be laws in the special sciences, like thermodynamics, chemistry, biology, geology, meteorology, psychology, economics, and so on. There are also objective probabilities and regularities involving them that occur in these sciences. More generally, there are theories in these sciences that enable the construction of causal patterns. Some may prefer not to call these laws, since they typically have exceptions, but I will continue to use the term. I will not provide anything close to a complete account of special science laws and probabilities here, but I want to say something about how they fit into the PDA.

Special science laws describe regularities that are formulated in languages that typically involve terms referring to macroscopic properties and entities and, unlike the laws of fundamental physics, specify causal relations among properties and events. Many philosophers have been puzzled about how there can be such laws given the enormous complexity of the distribution of physically fundamental entities and properties that conform to fundamental physical laws. Jerry Fodor expresses his puzzlement by first pointing out that:

> The very existence of the special sciences testifies to reliable macro-level regularities that are realized by mechanisms whose physical substance is quite typically heterogeneous.... Damn near everything we know about the world suggests that unimaginably complicated to-ings and fro-ings of bits and pieces at the extreme micro-level manage somehow to converge on stable macrolevel properties.[1]

[1] Fodor (1998) 160.

Fodor finds it, as he says, "molto mysterioso," that the motions of particles (and fields) to-ing and fro-ing in accordance with F = ma (or whatever the fundamental dynamical laws prove to be) end up converging on the regularities of the special sciences. How do the particles that, for example, constitute an economy "know" that their trajectories are required (ceteris paribus) to enforce, for example, Gresham's law in economics?

Fodor says that he doesn't expect to discover the answer to how there can be special sciences until the day before he finds out "why there is anything at all." While I doubt that any one mortal will ever adequately answer the second question, I think the PDA together with the actual fundamental laws of physics can provide an illuminating account of what special science laws are and how there can be special sciences. This should help alleviate Fodor's puzzlement as I will explain.

Recall that the PDA characterizes fundamental laws of physics in terms of a package that includes space-time, fundamental ontology, fundamental laws, and relevant macroscopic facts, including those specified by the special sciences. Law-determining candidates for the best system of the world are evaluated not only in terms of how well they systematize fundamental physics, but also in terms of how well they systematize facts involving macroscopic properties found in the special sciences and special science laws and how well they unify them with physics. For example, the fact that atomic theory can account for chemical elements, compounds, and regularities of chemical combination count in its favor. The fact that various chemical processes can account for biological processes and so on are parts of a unified system. At the foundation of the system are quantum mechanics and statistical mechanics. This is part of why quantum field theory is taken to be a candidate for a fundamental physical theory. I will soon explain why statistical mechanics is also foundational.

What makes a regularity a special science law? My view is that the laws of a special science systematize and unify events characterized in the vocabulary of that science much in the same way that laws of physics systematize events characterized in the vocabulary of physics. This is a page from the Humean book. In this I am following Callender and

Cohen and Schrenk.[2] In addition, the PDA also requires that special science systems themselves can be unified into a system constrained by the laws and ontology of fundamental physics. In this I differ from Callender and Cohen, who do not emphasize this unity.

Given the truths specified in the vocabulary of a special science, the theorems that are best unified by a system that balances informativeness, simplicity, and other scientific virtues are candidates for laws of the special science. These laws are typically probabilistic and/or include a ceteris paribus qualifier. Generalizations that include a ceteris paribus qualifier provide information by saying that a generalization holds with certain exceptions. According to Pietroski and Rey's important account, the exceptions are explained by their relations to independently specifiable factors.[3] These factors are typically subsumed under further laws of the special science or other sciences.

An example of a special science law of biology is Mendel's law of independent assortment which states that genes do not influence each other regarding the sorting of alleles into gametes, ceteris paribus. It is a law of genetics because of its role in a simple system that unifies an enormous amount of genetic information. The ceteris paribus qualifier is there because there are exceptions, but these exceptions are rare and can be explained by events that fall under other laws of biology or laws of other sciences.

How are the special sciences related to one another and to the laws and truths of fundamental physics? Each special science is autonomous in the sense that they are primarily pursued independently of one another and fundamental physics. Economists needn't learn quantum mechanics (their own subject is sufficiently confusing) although they may sometimes need to pay attention to psychology or meteorology. But the special sciences are not metaphysically autonomous. The laws of the various sciences must be compatible with each other and the laws of physics. Their probabilities and counterfactuals cannot conflict. The

---

[2] Cohen and Callender (2009) and Schrenk (2009) develop best systems accounts of special science laws. Schrenk shows how to provide truths conditions for ceteris paribus laws within such systems.

[3] Pietroski and Rey (1995).

truths of the special sciences supervene on fundamental physical facts. This is a fundamental tenet of physicalism. It is implied by the nomological/causal completeness of physics and the fact that special science phenomena have physical effects.[4]

This means that the laws and probabilities of physics unify laws and probabilities of the special sciences. If the laws of a special science entail $P(B/A) = x$ where A and B are propositions expressed in the vocabulary of a special science, and A* and B* are the same propositions as A and B, but expressed in the vocabulary of physics, then it had better be that $P(B/A) = P(B*/A*)$. If not, then the two sciences would be advising conflicting credences. Of course, in practice if such conflicts were to arise as a science develops, they typically would not be noticed. The reason is that the propositions of special sciences expressed in the language of physics are enormously complicated. But if a conflict is discovered, it needs to be resolved. Usually, the resolution will proceed by reformulating the special science law by adding further conditions which are required for it to hold.[5]

Physics is the father of the sciences keeping all their laws in line. But one of the special sciences is in a sense the mother of all the others. It plays a special role in unifying all of them with each other and with physics. That science is thermodynamics. Thermodynamics includes regularities that concern the transfer of energy and that relate macroscopic quantities like energy, mass density, pressure, temperature, radiation, frequency, and entropy. Since the regularities of the special sciences also involve processes that transfer energy, their phenomena are also covered by thermodynamics. Thermodynamics relates to fundamental physics in a way that I will shortly describe. As a consequence, the statistical mechanical probabilities that account for thermodynamics also provide the probabilities that appear in the special sciences. In this way,

---

[4] See Loewer () and Papineau () for arguments that the nomological/causal closure of physics implies that any phenomena that has physical consequences supervenes on fundamental physics.

[5] A famous case in which the special sciences conflicted is when geology indicated an age of the earth that was much less than natural selection seemed to require for the development of life. The conflict was resolved when the geological estimated age of the earth was revised based on considerations supplied by more fundamental physics.

statistical mechanics provides a probability map that unifies the special sciences with each other and with fundamental physics.[6]

The central law of thermodynamics is "the second law," which in its original form says that in the transfer of energy entropy never decreases and typically increases. Entropy is a measure of the energy in a system that is not available for work. Statistical mechanics accounts for how fundamental physics grounds thermodynamic quantities and laws. One way of formulating statistical mechanics, due originally to Boltzmann, is that there is a probability distribution over the possible initial conditions of the universe compatible with the hypothesis that the entropy of the universe at the time of the big bang was very small. When these are added to the fundamental dynamical laws, they yield a probability distribution over all contingent propositions. David Albert and I call this "the Mentaculus."[7] It is the kind of probability map of the universe mentioned in the previous chapter on chance.

It has been persuasively argued by Albert and others that the Mentaculus entails a probabilistic version of the second law that applies to the entire universe and to its approximately isolated subsystems. It also provides the basis for an account of counterfactuals and causation. This enables it to account for special science laws that involve causal relations and explanations. Explanations in the special sciences are ultimately based on the conditional probabilities entailed by the Mentaculus. If a special science law, causal relation, or explanation entails a conditional probability or counterfactual then that conditional probability and counterfactual must agree with the Mentaculus. It may seem astonishing that the laws and explanations of a special science like psychology, to the extent they are correct, are entailed by the Mentaculus, but this is required to avoid their being in conflict with physics. Defenses of these claims are found in Loewer (forthcoming b).[8]

The Mentaculus probability map is depicted by the diagram below.

---

[6] This claim is elaborated and defended in Loewer (forthcoming b).
[7] The name derives from the Coen Bros movie *A Serious Man* where a minor character refers to a book he is writing as "a probability map of the universe" and calls it "the Mentaculus."
[8] Mike Hicks in "Democracy for Laws" (2017) describes an account of laws which defends a similar account of how special science laws are related to laws of fundamental physics.

SPECIAL SCIENCE LAWS AND THE PDA 121

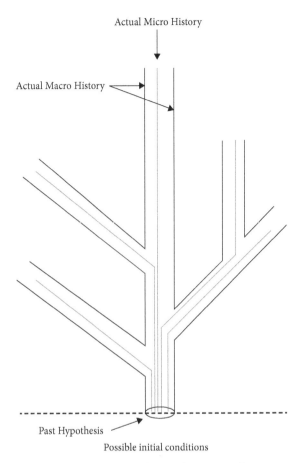

Probability Map of the Universe according to the Mentaculus

The thin lines represent possible micro-histories, and the red line represents the actual micro-history. The circle represents the special low entropy macro-state postulated by PH. Of course, the circle occupies only a tiny portion of the space of nomologically possible initial conditions. The cylinders represent macro-histories, and the blue cylinder represents the actual macro-history. All the micro-histories that initiate in the circle satisfy the low entropy initial condition and almost all evolve deterministically towards a state of maximum entropy. Although the evolutions of micro-histories are governed by deterministic laws, macro-histories evolve probabilistically. That is, the macro-state at t, together with the dynamical laws, don't determine a unique evolution, but the probability distribution specifies the probabilities of the histories that realize that macro-state.[9] The branching of the cylinders represents the indeterministic evolution of macro-histories. It is possible for macro-histories to converge but this is enormously unlikely.

The probabilities and probabilistic regularities of the special sciences arise from conditionalizing on the Mentaculus distribution. For example, that the probability of a coin when flipped on earth in a certain situation is 0.5 derives from conditionalizing on the macro conditions of the coin's flipping (that it is on earth, with certain gravitational field, certain atmospheric conditions, with a certain approximate force, and so on). Exactly how the coin lands if it is properly flipped is sensitive to exactly how it is flipped and that depends on the random behavior of human beings and their environment which ultimately is due to statistical mechanics.

The probabilistic map provided by the Mentaculus entails that over time certain probabilistic regularities become stabilized so that they hold under a variety of conditions. This is described, for example, by evolutionary history. At some time, and for a period of time, it is likely that, due to natural selection, the relative sizes of predator and prey animal populations in a region become stabilized. The Mentaculus provides the probabilities that underlie this process. My claim is that the probabilities in the Mentaculus explain why the to-ing and fro-ing of fundamental particles converge on special science generalizations like this one. That some of these regularities are counted as laws are due to their fitting into a system characterized in the special science vocabulary. This should help alleviate Fodor's puzzlement of how there can be special science laws.

The primary novel idea of the PDA is that it dispenses with Lewis's perfectly natural properties and the BSA claim that the optimal systematization of *their* distribution determines laws. The PDA replaces this with the claim that candidate optimal systems aim to systematize truths

---

[9] The probability assignment over initial conditions together with the deterministic laws and the supervenience principles connecting macro- to micro-states determines a probability distribution over propositions characterizing macro-states. So, for example, it assigns a probability to the proposition that a particular ice cube in warm water will melt in the next hour.

expressed in macroscopic and special science vocabulary. Doing this requires their adding further ontology and vocabulary, including the ontology and vocabulary of fundamental physics. It is this entire package that comprises a system that determines which propositions are laws and which expressions refer to fundamental properties and quantities. This chapter has shown how the special sciences are a crucial component of this package.

10
Realism, Relativism, and Reference

There is a worry that the BSA and even more so the PDA may be incompatible with scientific realism and metaphysical realism and, worse, be committed to a pernicious type of relativism. The worry is due to the fact that both accounts say that laws and, in the case of PDA, fundamental ontology/properties and space-time, depend in part on what systems best satisfy criteria fashioned by human scientific practice. These values include simplicity, scientific informativeness, fit, and, in the case of the PDA, systematizing scientifically relevant macroscopic properties and special science laws. These macroscopic properties and laws are results of the practice of science which itself depends on human interests. If the laws and fundamental ontology are partly human creations, doesn't that provide reasons to doubt the realist credentials of the BSA and PDA? In this chapter, I will argue that they are compatible with scientific realism and that, although they make laws and, in the case of the PDA, fundamental ontology and properties partly dependent on scientific practice, this poses no problem for defensible metaphysical realism. While it might not be what some require of metaphysical realism, it is realism enough for science to be objective. While the PDA does make laws and fundamental ontology relative to scientific criteria, this relativism is not problematic and, in fact, allows it to avoid the problem Dasgupta raised about the value of naturalness without succumbing to the extreme relativism he seems to advocate.[1]

Scientific realism is usually understood as including the following three claims:

---

[1] There is an excellent discussion of the tension between Humean accounts of laws like the BSA and the PDA by Bixin Guo (forthcoming). Much of what I say is in response to her paper.

(i) Scientists aim to produce true theories and explanations.
(ii) There is evidence that some theories that have been produced are true or approximately true, even in what they say about unobservable phenomena.
(iii) Theories that satisfy all epistemic criteria except truth including all empirical tests may yet be false.

It is clear that both the BSA and the PDA are compatible with (i) and (ii). If epistemic criteria consist in being compatible with all nomologically possible observations and satisfying scientific criteria except for truth about unobservable propositions then they are both compatible with (iii) as well. Theories that are empirically equivalent can have different fundamental ontologies/properties and different laws and so both cannot be true, but only one can specify the laws and fundamental ontology. This is true for both the BSA and PDA.

Scientific realists also believe that there are laws of nature, causation, counterfactuals, dispositions, and other nomological modalities since these are essential to scientific explanation. But this poses no problem to the compatibility of scientific realism with the BSA and the PDA as long as laws are understood as the BSA and PDA understand them. Of course, the BSA and the PDA reject that there are governing laws and laws derived from the operations of powers and any nomological modalities that depend on them. But scientific realism is not committed to them. These are philosophical views about the nature of laws. The conclusion is that there is no incompatibility between scientific realism and either the BSA or the PDA as long as they provide adequate accounts of nomological modalities.

What about metaphysical realism? The best way to approach this issue is by way of discussing Lewis's worry that his BSA might succumb to a disease he called "ratbag idealism." According to the dictionary of Australian and New Zealand slang calling something "ratbag" is to say that it is rascally or is in some way despicable and unpleasant. For Lewis, ratbag idealism is the view that the laws depend on us—on our interests, psychology, and the history of science in a way that undermines their objectivity.[2] He worried that ratbag idealism threatens the BSA's realist credentials because criteria for evaluating candidates for best system like simplicity and informativeness depend on scientific practice and human psychology. If this is a problem with the BSA, it seems even worse for the PDA, since the PDA adds further criteria to the evaluation of candidates for best system that have been shaped by the history of science, and it applies not only to laws but also to scientifically fundamental ontology, properties, and fundamental space-time.

Here is Lewis on ratbag idealism:

> The worst problem about the best-system analysis is that when we ask where the standards of simplicity and strength and balance come from, the answer may seem to be that they come from us. Now, some ratbag idealist might say that if we don't like the misfortunes that the laws of nature visit upon us, we can change the laws—in fact, we can make them always have been different—just by changing the way we think! (Talk about the power of positive thinking.) It would be very bad if my analysis endorsed such lunacy. I used to think rigidification came to the rescue: in talking about what the laws would be if we changed our thinking, we use not our hypothetical new standards of simplicity, strength, and balance, but rather our actual and present standards. But now I think that is a cosmetic remedy only. It doesn't make the problem go away; it only makes it harder to state. The real answer lies elsewhere: if nature is kind to us, the problem needn't arise. I suppose our standards of simplicity and strength and balance are only partly a matter of psychology. It's not because of how we happen to think that a linear function is simpler than a quartic or a step function; it's not because of how we happen to think that a shorter alternation of prenex quantifiers is simpler than a longer one; and so on. Maybe some of the exchange rates between aspects of simplicity, etc., are a psychological matter, but not just anything goes. If nature is kind, the best system will be robustly best—so far ahead of its rivals

---

[2] According to Gordon Belot (2021): "a ratbag idealist is someone who takes some pretheoretically fundamental aspect of the world—its causal structure, or its laws, or its spatiotemporal geometry—to depend on human cognitive constitution."

that it will come out first under any standards of simplicity and strength and balance. We have no guarantee that nature is kind in this way, but no evidence that it isn't. It's a reasonable hope.[3]

I agree with Lewis that there is no guarantee that nature has either a Lewisian or a PDA best system. The enterprise of physics assumes that there is a best system and seeks to find one. So far, this assumption seems to have paid off. It is true that, on both accounts of laws, which propositions express laws is relative to human interests, psychology, and so on, to the extent that the criteria used to evaluate candidates for best system are relative to us. By making fundamental properties/ontology and space-time, as well as laws, part of the package, the PDA seems to also make them hostage to criteria developed during the history of science and to that extent also relative to human interests and practices. Adding ontology and space-time to the package may make it easier for nature to be kind, since it may make it more likely that there exists a PDA best systematization, but it also seems to make them relative to human interests.

So, it must be granted that there is relativism in the PDA. Since a best system is supposed to explain or explain away the behaviors of macroscopic objects as characterized in the human languages with which physics begins, as well as satisfy criteria that have been developed over the history of scientific practice, being a best system is relative to human beings and human scientists. But this does not diminish the fact that, given scientific criteria, it is an objective matter whether the world has a best system and which, if any, systems qualify as best. The truth value of the system is an objective matter completely independent of human interests. What *does* depend on the interests of human science is which of those propositions count as laws and what properties/ontology count as scientifically fundamental and the nature of the fundamental space-time arena of the universe. Which truths are laws, which predicates denote fundamental properties, and the nature of fundamental space time is a consequence of the best systematization and what is a best

---

[3] Lewis (1994) 479.

systematization is relative to scientific interests.[4] Does this undermine the BSA's and PDA's realist credentials? I don't believe it does.

Gordon Belot says that the test of whether an account of laws is ratbag idealist is whether there could be a group of intelligent beings whose scientific practices and criteria for optimal systems differ from human science so significantly as to result in different propositions being laws.[5] Suppose there are extra-terrestrials whose cognitive capacities so far outstrip humans and whose scientific practices are so different that they count as laws propositions that when expressed in human languages are much too complex to be laws for humans. Note that the propositions they consider laws must be true. Perhaps their senses can detect complex non-local property instantiations and they consider such properties simple so that they play a role like ordinary macro propositions play in the PDA. Laws for them would differ from laws for us. Perhaps they use their laws to make predictions or build devices that operate in ways that we would find miraculous.

Doesn't this possibility show that the BSA and the PDA succumb to ratbag idealism? If it does, I will argue that it is harmless. It does not compromise realism or present a problem for either view. We and these alien creatures may differ on which regularities count as laws, and which properties count as fundamental, but we need not disagree which propositions are true as long as the propositions don't implicate laws or what is fundamental. They may count some propositions as laws that we think are merely accidental truths, and this may enable them to make predictions and provide explanations beyond what we can do. But if our laws are, as they may well be, dynamical, deterministic, and complete, then if we were to encounter such beings, we should think that they have discovered initial conditions of the universe which they can characterize as laws even though these conditions are too complex for us to count them as laws in our language. Given the usual statistical mechanical probability distribution

---

[4] It is important to recognize that this does not mean that macroscopic properties or the structure of familiar 3-dimensional space is dependent on human cognitive/scientific interests. They are aspects of "the manifest image." It does mean that these can be accounted for in terms of fundamental laws, properties, and space time and the fact that these are fundamental does depend on human/scientific cognitive interests. These are aspects of "the scientific image" created by science.

[5] My discussion owes a lot to Belot's (2022).

over initial conditions, the existence of such beings is very unlikely. We would think that their success was due to "luck."

If they were to persist in their "miraculous" predictions, we should come to see it as not miraculous. In fact, we should believe that there is a characterization of the initial conditions of our universe that explains their success but is too complex expressed in the language of our physics for us to count as lawful. If we can find a way of characterizing these conditions that meets our scientific criteria, we can add this characterization to the best system. If we can't find such a characterization, we could even expand scientific criteria to include reference to these extra-terrestrials. Adding such a law to the fundamental laws involves adding a law very unlike anything else seen in the history of science. The Past Hypothesis is also a law expressed in macroscopic terms, but it involves nothing so unusual as reference to extra-terrestrials.

My conclusion is that given what we know from our current physics, it is very unlikely that there are intelligent creatures who, even though their macroscopic concepts and special sciences differ from ours, develop a fundamental physics that is very different from ours. This should dispel the threat of ratbag idealism.

A different threat to the PDA's realist credentials comes from Lewis's theory of reference magnetism. According to it, perfectly natural properties play a central role in accounting for the references of terms and, in particular, the predicates that appear in candidates for fundamental systems. Lewis uses this account of reference to respond to Putnam's model theoretic argument against metaphysical realism. If this response is required, then, contrary to the PDA, perfectly natural properties would also play a central role in the account of laws. My attempt to do without them would come to naught.

Putnam argues that metaphysical realism is false, since it requires a magical account of reference. He seems to think that the only non-magical account of reference that could possibly be compatible with metaphysical realism is the view that the references of terms are required to make the theory true.[6] If this were so, then an ideal scientific theory,

---

[6] Why he thinks this is not completely clear. Apparently, his view is that metaphysical realism involves the view that if metaphysical realism as he understands it is true then what is going on in one's mind provides no constraints on what is going on outside of one's mind. This

i.e. a consistent theory that meets all scientific criteria, including empirical accuracy, simplicity, and so on, must be true, since it will have a true interpretation as long as the world contains sufficiently many items. But scientific realism is usually understood as including the claim that even an ideal theory may be false. Putnam is claiming that metaphysical realism undermines scientific realism, an astonishing result. An obvious reply is that there are other conditions than making a theory true that determine the references of its terms. If this is so, then the theory may be false. But Putnam argues that, given metaphysical realism, adding further conditions on what counts as a correct interpretation doesn't help, since it is just adding "more theory" and so it is subject to the same argument.

Lewis responds to Putnam's argument with his account of reference magnetism. This account says that the references of the predicates of our language are the most natural properties that satisfy whatever other constraints are imposed by a naturalistic account of reference. Examples of other constraints are that there are certain causal, informational, or counterfactual relations between uses of the predicate and its reference. Naturalistic accounts of reference developed by Dretske, Fodor, Millikan, and others in the 1980s go some distance in characterizing the references of terms, but do not provide sufficient conditions to select a term's reference from among many candidates. Lewis proposed adding the condition that from among the admissible referents the one or ones with the greatest degree of naturalness are the referents.[7] He claims that this theory of reference is not just "adding more theory" but is rather a presupposition of our concepts and terms having reference at all. The idea is that naturalness acts as a kind of magnet attracting a concept or term to its reference. The threat to the PDA is that if Lewis's reply is the only way to rebut Putnam's argument, then perfectly natural properties

would entail, for example, that a brain in a vat on the moon could have exactly the same thoughts as a person watching an NBA basketball game as long as their internal brain states are the same, even though the brain in a vat had never encountered basketball games, etc. Of course, Putnam is famous for arguing that this isn't really possible, saying, as he put it, "meanings ain't in the head." Lewis seem to think that meanings are in the head but that features of the world outside of the head, namely naturalness, constrain the meanings of terms in the head.

[7] See Loewer (2017) for a survey of naturalistic accounts of reference.

and degrees of naturalness are needed to save realism, and my attempt to do without perfect naturalness will have failed.

I am not confident that Putnam's argument is correct or, if it is, that Lewis's reply rebuts it. But I do maintain that if Lewis's reply is effective against Putnam's argument, then the scientifically fundamental properties as characterized by the PDA can replace his natural properties in that argument. My proposal is that the meanings and references of the concepts and terms with which scientific theorizing begins are taken as given and that they restrict interpretations of the rest of our theory so as to avoid Putnam's argument. Alternative interpretations will either get these wrong or lead to complications that violate the PDA's criteria.

Lewis's magnetism account of reference truly is magical. He provides no reason to believe that his perfectly natural properties are reference magnets. On the other hand, the view that naturalness, characterized in terms of scientifically fundamental properties, plays a role in fixing the references of terms and concepts is not implausible. Scientifically fundamental properties are characterized by the role they play in a theory that accounts for macroscopic phenomena characterized in the languages with which science begins. It is not surprising then that scientifically fundamental properties and naturalness defined in terms of them should be involved in a naturalistic account of the reference of macroscopic scientific terms.

There is a kind of circularity in the preceding account of reference. The references of the macroscopic predicates with which science begins are taken as given. An ideal theory that optimally satisfies scientific criteria then determines which properties/quantities are scientifically fundamental. This determines which properties are perfectly scientifically natural and the degree to which other properties are scientifically natural. Together with a naturalistic account of reference of the sort that Dretske, Fodor, Millikan, et al. were attempting to find, the referents of all predicates, including the macroscopic ones with which science begins, are determined. Therein lies the circularity. The more fundamental theories will often correct some of the beliefs with which science begins, leading to adjustments in the references of some predicates. It is an assumption of the entire project of science that this will all hang together.

11
Reprise and Conclusion

The primary goals of this book have been to argue that Lewis's Humean BSA of laws is superior to non-Humean accounts, and then show that the Package Deal Account is a major improvement over Lewis's BSA. The BSA develops the Humean idea that what makes a proposition or equation a law is its role in systematizing the universe and dispenses with the idea that laws "govern" the universe by producing it or constraining it. The PDA improves the BSA by dispensing with metaphysically prior perfectly natural properties as the universe's fundamental properties. Doing this advances the project of naturalizing metaphysics. This final chapter summarizes the various features of the PDA that make this the case and brings my discussion to a conclusion.

The PDA earns its Humean credentials by rejecting the non-Humean view that laws "govern," "constrain," or are the product of fundamental necessary connections, in favor of the view that laws systematize and unify. It preserves the core idea of the BSA that laws supervene on the distribution of instantiations of scientifically fundamental properties and are determined by axioms that optimally systematize this distribution, but by making the laws and properties parts of a package it doesn't make either prior to the other. It is in virtue of their places in a systematization that true propositions qualify as lawful and certain properties qualify as scientifically fundamental.

Both non-Humean and Humean accounts of laws have theological roots. Non-Humean accounts derive from the idea that laws are the Deity's means of governing. Both the BSA and the PDA promote the systematizing aspect of laws, which also has a theological basis. But unlike Lewis's BSA, the PDA does not rely on the world being, as Putnam says, "ready-made" with a unique division into metaphysically fundamental properties that physics has the job of revealing. The "ready-made" structure of the world can be seen as a remnant of the theological origin of the concept of laws, too. It is the way the Deity constructed the world. Instead, the PDA allows that there may be multiple ways of counting properties as "scientifically fundamental," in virtue of being included in packages that optimally satisfy criteria including systematizing the macroscopic phenomena with which science begins.

By bringing macroscopic and special science properties into the package that determines fundamental laws, the PDA provides a place for special science laws and accounts for the connections between special science properties and laws and scientifically fundamental properties and laws. It also allows for developing the BSA account of objective chance to include probabilistic laws.

In Chapter 5, I discussed five problems that result from basing the BSA on perfectly natural properties. The five are:

(i) There are properties considered to be fundamental in theories proposed in physics that don't satisfy all of Lewis's conditions of perfect naturalness.
(ii) Fundamental physics, while interested in laws, doesn't seem interested in finding a unique set of fundamental properties. Equivalent scientifically best systems can be formulated which take different properties to be fundamental.
(iii) Perfectly natural properties are quiddities and so are unknowable by science, but there are fundamental properties in physical theories that seem to have at least some of their nomological roles necessarily.
(iv) It leads to the mismatch problem.
(v) And it leads to the missing value problem.

The PDA handles all five.

According to the PDA, there need not be a unique collection of scientifically fundamental properties. Different TOEs that differ in their basic predicates, and so differ in what properties/quantities and laws they count as fundamental, yet may still be true. This is illustrated by Newtonian, Lagrangian, and Hamiltonian formulations of classical mechanical laws and alternative versions of quantum mechanics that are equivalent in all ways, save what they count as fundamental.

The BSA is committed to saying that the properties of at most one of these theories are perfectly natural, but the PDA can count more than one as scientifically fundamental.

Unlike the BSA, the PDA has no difficulty accommodating quantum mechanics, string theory, or the QFT representation of quantities (e.g. quark color) as fiber bundles, since it makes no metaphysical commitment to the structure of the fundamental space-time or to whether the quantities referred to by predicates and function symbols in a TOE are intrinsic or categorical. It leaves it to physics to decide on these, if it has reason to. The PDA can accommodate proposals for fundamental theories with space-times with many dimensions and exotic geometries or with no fundamental space-times at all. Unlike the BSA, the PDA does not assume a metaphysically given space-time structure, but rather treats the arena in which fundamental ontology and quantities are instantiated as elements of a package. It has no difficulty accommodating proposed loop quantum gravity and other quantum gravity theories that treat space-time as emergent from something more fundamental.[1]

One issue that has received discussion in philosophy of physics is whether symmetries exhibited by dynamical laws determine space-time geometry or the other way round. Since space-time or the arena that plays the role of space-time comes together with the laws and fundamental ontology, the issue of whether the symmetries in the laws partly determine space-time structure or space-time partly determines the symmetries in the laws is resolved. Neither has priority over the other since they come together as components of a single package.[2]

One of the most interesting features of the PDA is that it is non-committal about whether scientifically fundamental properties are categorical, or dispositional, or some combination. This is due to the fact that the properties/quantities referred to in a TOE can generally be understood either as categorical or dispositional. Consider, for example, that in Newtonian mechanics the predicate "mass" can be interpreted as referring to either a categorical or to a dispositional quantity. If the

---

[1] Lam and Wüthrich (2023).
[2] In this way the PDA implements D.C. Williams idea that what makes one of the dimensions the temporal dimension is its place in dynamical laws (Williams 1951).

latter, then mass necessarily satisfies the Newtonian laws. But these necessary connections should not offend Humean sensibilities, since the predicates and the properties they refer to are specified by a scientifically best system. The necessary connections arise because they are parts of the best systematization. They do not determine the laws. Instead, the systematization determines both the laws and the ontology.

The fact that the PDA allows for scientifically fundamental properties that are categorical and also scientifically fundamental properties that are dispositional does not solve the problem of whether metaphysically *fundamental* properties are categorical or dispositional. This is not a problem that needs to be solved by an account of scientifically fundamental properties and laws. The PDA doesn't say anything about what properties are *metaphysically* fundamental or even if there are metaphysically fundamental properties. At its most metaphysically fundamental level, reality might be a whole that possesses or can sustain multiple structures. The criteria that physics has developed for assessing candidate optimal systems determines which of those structures are *scientifically fundamental*.

The mismatch problem cannot arise on the PDA, since the scientifically fundamental predicates and the fundamental theory are determined together with each other as a package. While there is no guarantee that science will ever discover a TOE for the world, the PDA does guarantee that if one is discovered, its axioms will be fundamental laws, the generalizations it entails will be nomologically necessary, and its basic predicates will be scientifically fundamental.

The PDA also points towards a solution to Dasgupta's missing value problem without sliding into relativism. Dasgupta asked why systematizing the distribution of instantiations of Lewis's perfectly natural properties is more valuable than systematizing the distribution of natural*, or natural**, or... properties. This is a way of asking for an explanation of why systematizing perfectly natural properties are of value to physics at all. The PDA answers the challenge for scientifically fundamental properties by providing an account of why it is valuable to systematize the macroscopic mosaic. It is valuable because there is value in knowing how macroscopic objects move. Our lives may depend on and have been improved by this knowledge. Furthermore, the reason it is of value to discover scientifically fundamental properties is that they are components of a TOE that optimally satisfies criteria that have been developed and shaped by science. These criteria are designed to achieve the aim of optimally organizing, predicting, and explaining truths, initially about macroscopic phenomena and then about all the phenomena in its domain. It is evident that discovery of lawful macroscopic regularities and more fundamental laws that underly them is epistemically valuable for the community of scientists and that it has consequences that are valuable more generally.

I argued in Chapter 4 that defenders of the BSA can rebut the worry that Humean laws can't explain due to the fact that the laws depend on the Humean mosaic. But the worry doesn't even arise on the PDA since, according to it, laws and properties are part of a package that optimally scientifically systematizes the world. While the laws supervene on the distribution of fundamental property instantiations they are not determined by that distribution. The worry that laws can't underly explanations of the mosaic because the mosaic explains the laws doesn't get a foothold. On the PDA, as on the BSA, laws explain by unifying and systematizing.

A notable feature of the PDA is that it is not subject to Lazarovici's argument that Lewis's Humeanism entails that systematizable worlds are atypical. Recall that this argument is that every combination and recombination of perfectly natural properties in a possible space-time composes a possible world, and since systematizable worlds are a rarity among these, it would be incredible that our world is systematizable. Since the PDA does not assume a Humean metaphysics of fundamental categorical properties that can be arbitrarily combined to form possible worlds, it is not subject to this argument. In fact, the PDA is not committed to any view about metaphysical possibility, other than that systematizable worlds are possible and there are possible worlds compatible with the laws of any systematization of the actual world. This is sufficient to provide truth conditions for statements about physical necessity and counterfactuals. But although possible worlds are invoked in specifying truth conditions, there is no commitment to any metaphysical view about what possible worlds are. And, as observed earlier, on Humean accounts of counterfactuals, like Lewis's and, the truth makers of counterfactuals can be understood as consisting solely of facts about the actual world that don't involve fundamental necessity.

The PDA also helps alleviate the worry that the atypicality of systematizable Humean mosaics results in problems concerning induction that are worse than those encountered by other accounts of laws. Recall Ismael's worry that on Humeanism finding out about one portion of the mosaic gives us no reason to believe anything about other portions. Because the PDA rejects perfectly natural properties and the recombination principle, these problems do not arise based on its metaphysical commitments as they do on Lewis's account. Since scientifically natural properties may have necessary connections to one another, they may not be arbitrarily combinable. This does not mean that on the PDA we should not be astonished that the world is systematizable, to the extent that it is, or that the PDA solves the problem of justifying induction. But on every account of laws, we should be astonished that the world is systematizable, and no account solves the problem of justifying induction.

I haven't said much about what, if anything, underlies a best system's distribution of scientifically fundamental properties. The metaphysics that is best suited to the PDA is a kind of metaphysical holism.[3] On this account, what is metaphysically fundamental is a reality that may support multiple structures, only some of which are scientifically fundamental. An optimal scientific system reflects a structure that includes a space-time or an arena that plays a similar role, an ontology with fundamental properties and a system of laws. It cuts reality into parts that may include a space-time, fields, individuals, wave functions, entanglement relations, and so on. There may be many ways this can be done. Reality may endorse multiple optimal scientific structures and structures which are not scientific but satisfy other criteria. An optimal scientific structure

---

[3] The monism proposed here is similar to what Schaffer (2010) call "priority monism" since it holds that the one cosmos is prior to whatever ontology an optimal scientific systematization may include. But it also is different from his version of priority monism, since his version concerns the ontology that he takes to be the unique scientific system of the world. He thinks it may be quantum mechanical entanglement relations that unify the cosmos, but quantum mechanics allows systems that are not entangled with one another, and cosmology suggests the possibility of multiple universes that are parts of a mother universe, and which are not entangled with each other. In any case, on my version of monism, reality/the cosmos is prior to the ontology and properties that are scientifically fundamental.

is of special interest to us because it is connected to and unifies macroscopic facts and provides the machinery for scientific explanations.

The metaphysical story just told may be fanciful. It is also compatible with the PDA that there is a unique metaphysically fundamental structure consisting of space-time filled with instantiations of Lewis's perfectly natural properties. But if there are such properties, they need not be scientifically fundamental, and it is not the systematization of their distribution that determines the laws, according to the PDA.[4] It is compatible with the PDA that there are perfectly natural properties, and that they play some of the roles Lewis assigns to them in his metaphysics. The main point is that the PDA is not committed to any account about what, if anything, is metaphysically fundamental except that what is metaphysically fundamental supports possible scientifically fundamental structures and that it determines the truth values of all propositions.[5]

On the PDA, reality makes propositions true. An optimal systematization determines which propositions count as laws and which properties count as fundamental. There is no need for governing laws, primitive powers, or a Deity to enforce regularities and no need for perfectly natural properties to support its laws and ontology.

Lewis's metaphysical viewpoint is that of a god surveying all of a ready-made reality to find its best systematization. Reality comes with its space-time structure and distribution of perfectly natural properties/quantities. It is the job of physics to find the laws by finding its best systematization. Lewis's realism about laws imagines a transcendental perspective, but the PDA is more an account of laws, fundamental properties/ontology, and fundamental space-time from within the activity of science. As Quine says:

---

[4] A similar idea is developed in Bhogal and Perry (2023).

[5] Sometimes Lewis writes as though he thinks that the perfectly natural properties just are the properties that science would find to be fundamental. This would make his view more like the PDA. But this way of understanding Lewis isn't compatible with the roles that perfectly natural properties play in his account. He posits that they have certain features, for example that they are categorical, instantiated at points, are reference magnet, that there is a unique collection of them, etc., but nothing guarantees that scientifically fundamental properties have these features, and in fact it seems to have been already decided that they don't. I think it is clear that he posits them as a metaphysical presupposition and then trusts that science will hit upon them. This footnote was prompted by a conversation with Veronica Gomez.

> It is understandable [...] that the philosopher should seek a transcendental vantage point, outside the world that imprisons [the] natural scientist and mathematician. He would make himself independent of the conceptual scheme which it is his task to study and revise. "Give me που στω [a place to stand]", Archimedes said, "and I will move the world." However, there is no such cosmic exile. [...] The philosopher is in the position rather, as Neurath says, "of a mariner who must rebuild his ship on the open sea".[6]

The PDA takes Quine's perspective seriously. It replaces the view that the goal of science is to discover metaphysically given perfectly natural properties with the view that scientifically fundamental properties, laws, and space-time are the product of science working from within the world in conformity with principles developed within science.

Hawking asked, "What is it that breathes fire into the equations and makes a universe for them to describe?" The PDA's answer to his question is that the fire is the activity of science. Reality's structure provides the fuel, and science, by finding a systematization satisfying certain conditions developed during its history, ignites it to yield the world's fundamental ontology and its laws.

---

[6] W. V. Quine (in notes for *Sign and Object*, November 5, 1944) as quoted by Verhaegh (2018).

References

Albert, David (1992). *Quantum Mechanics and Experience*. Cambridge, MA: Harvard University Press.
Albert, David (2000). *Time and Chance*. Cambridge, MA: Harvard University Press.
Albert, David (2023). *A Guess at the Riddle*. Cambridge, MA: Harvard University Press.
Armstrong, David (1980). *What Is a Law of Nature?* Cambridge: Cambridge University Press.
Balog, Katalin (2012). "In Defense of the Phenomenal Concept Strategy." *Philosophy and Phenomenological Research* 84 (1): 1–23.
Barbour, Julian (1982). "Relational Concepts of Space and Time." *British Journal for the Philosophy of Science* 33 (3): 251–74.
Beebee, Helen (2000). "The Non-Governing Conception of Laws of Nature." *Philosophy and Phenomenological Research* 61 (3): 571–94.
Bell, E. T. (1945). *The Development of Mathematics*, 2nd edition, 145. New York: McGraw-Hill Book Company.
Bell, John S. (1987). *Speakable and Unspeakable in Quantum Mechanics*. Cambridge: Cambridge University Press.
Belot, G. (2022). "Ratbag Idealism." In Y. Ben-Menahem (ed.), *Rethinking the Concept of Law of Nature: Natural Order in the Light of Contemporary Science*, 1–20. Cham: Springer International Publishing.
Bhogal, Harjit (2020). "Nomothetic Explanation and Humeanism about Laws." In Karen Bennett and Dean Zimmerman (eds), *Oxford Studies in Metaphysics*, 12:164–202. Oxford: Oxford University Press.
Bhogal, Harjit, and Zee Perry (2023). "Humean Nomic Essentialism." *Nous* 57 (1): 81–99.
Bird, Alexander (2005). "The Dispositionalist Conception of Laws." *Foundations of Science* 10: 353–70.
Blanchard, Thomas (2023). "Best System Laws, Explanation and Unification." In Michael Townsen Hicks, Siegfried Jaag, and Christian Loew (eds), *Humean Laws for Human Agents*, 171–185. Oxford: Oxford University Press.
Carnap, Rudolf (1945). "On Inductive Logic." *Philosophy of Science* 12(2): 72–97.
Carroll, Sean (2010). *From Eternity to Here: The Quest for the Ultimate Theory of Time*. Boston: Dutton.
Carroll, S. M. (2022). "The Quantum Field Theory on which the Everyday World Supervenes." In Stavros Ioannidis, Gal Vishne, Meir Hemmo, Orly Shenker (eds), *Levels of Reality in Science and Philosophy: Re-Examining the Multi-Level Structure of Reality*, 27–46. Cham: Springer International Publishing.
Cartwright, Nancy (1983). *How the Laws of Physics Lie*. Oxford: Oxford University Press.
Cartwright, Nancy (1989). *Nature's Capacities and Their Measurement*. Oxford: Oxford University Press.
Cartwright, Nancy (1999). *The Dappled World: A Study of the Boundaries of Science*. Cambridge: Cambridge University Press.

Cartwright, Nancy (2005). "No God; No Laws." In E. Sindoni and S. Moriggi (eds), *Dio, la natura e le legge. God and the Laws of Nature*. Milan: Angelicum-Mondo X.

Chen, Eddy (2022). "The Wentaculus: Density Matrix Realism Meets the Arrow of Time." https://arxiv.org/abs/2211.03973.

Chen, Eddy Keming (2022). "Fundamental Nomic Vagueness." *Philosophical Review* 131 (1): 1–49.

Chen, Eddy, and Shelly Goldstein (2023). "Governing without a Fundamental Direction of Time: Minimal Primitivism about Laws of Nature." In Yemima Ben-Menahem (ed.), *Rethinking Laws of Nature*, 21–64. New York: Springer.

Clarke, Samuel (1738). "The Evidences of Natural and Revealed Religion." In *The Works of Samuel Clarke*, D.D., 2:698 (2 vols, London, 1738).

Coffey, Kevin (2014). "Theoretical Equivalence as Interpretative Equivalence." *The British Journal for the Philosophy of Science* 64 (4): 1–30.

Cohen, Jonathan, and Craig Callender (2009). "A Better Best System Account of Lawhood." *Philosophical Studies* 145 (1): 1–34.

Cournot, Antoine August (1843). *Exposition de la theorie des chances et des probabilities*. Paris: Librairie de L. Hachette.

Dasgupta, Shamik (2018). "Realism and the Absence of Value." *Philosophical Review* 127 (3): 279–322.

Demarest, Heather (2017). "Powerful Properties, Powerless Laws." In Jonathan D. Jacobs (ed.), *Causal Powers*, 38–53. Oxford: Oxford University Press.

Dorr, Cian (2019). "Natural Properties." In *Stanford Encyclopedia of Philosophy*. https://plato.stanford.edu/entries/natural-properties/ sec. 2.6

Dorst, Chris (2019). "Towards a Best Predictive System Account of Laws of Nature." *British Journal for the Philosophy of Science* 70 (3): 877–900.

Dorst, Chris (2020). "Why Do Laws Support Counterfactuals." *Erkenntnis* 87 (2): 545–66.

Dretske, Fred (1977). "Laws of Nature." *Philosophy of Science* 44: 248–68.

Eagle, Anthony (2004). "Twenty-One Arguments against Propensity Analyses of Probability." *Erkenntnis* 60: 371–416.

Earman, John (1986). *A Primer on Determinism*. Dordrecht: Reidel.

Elga, Adam (2004). "Infinitesimal Chances and the Laws of Nature." *Australasian Journal of Philosophy* 82 (1): 67–76.

Esfeld, Michael (2019). "Super Humeanism and Free will." *Synthese* 198 (7): 6245–58.

Esfeld, Michael, and Dirk-André Deckert (2017). *A Minimalist Ontology of the Natural World*. New York: Routledge.

Field, Hartry (2003). "Causation in a Physical World." In Micheal Loux and Dean Zimmerman, eds, *Oxford Handbook of Metaphysics*, 435–460. Oxford: Oxford University Press.

Filomeno, Aldo (2021). "Are Non-accidental Regularities a Cosmic Coincidence? Revisiting a Central Threat to Humean Laws?" *Synthese* 198: 5205–27.

Fodor, Jerry (1974). "Special Science or the Disunity of the Sciences as a Working Hypothesis." *Synthese* 28: 97–115.

Fodor, Jerry (1998). "Special Sciences: Still Autonomous After All These Years." In Jerry Fodor (ed.), *In Critical Condition*, 9–24. Cambridge, MA: MIT Press.

Foster, John (2004). *The Divine Lawmaker: Lectures on Induction, Laws of Nature, and the Existence of God*. Oxford: Oxford University Press.

French, Steven, and Kerry McKenzie (2012). "Thinking Outside the Toolbox: Towards a More Productive Engagement between Metaphysics and Philosophy of Physics." *European Journal of Analytic Philosophy* 8 (1): 42–59.

Friedman, Michael (1974). "Explanation and Scientific Understanding." *Journal of Philosophy* 71: 5–19.
Frigg, Roman (2008). "Chance in Boltzmannian Statistical Mechanics." *Philosophy of Science*, 75 (5): 670–681.
Fuchs, Christopher A., N. David Mermin, and Rüdiger Schack (2014). "An Introduction to QBism with an Application to the Locality of Quantum Mechanics." *American Journal of Physics* 82 (8): 749–54.
Guo, Bixin (forthcoming). "Can Humeans Be Scientific Realists?" (under review *BJPS*).
Hacking, Ian (1984). *The Emergence of Probability*. Orig. 1975. Cambridge: Cambridge University Press.
Hájek, Alan (1996). "'Mises Redux'—Redux: Fifteen Arguments against Finite Frequentism." *Erkenntnis* 45: 209–27.
Hajek, Alan (1997). "Fifteen Arguments against Finite Frequentism." *Probability, Dynamics and Causality* 45 (2/3): 209–27.
Hájek, Alan (2009). "Fifteen Arguments against Hypothetical Frequentism." *Erkenntnis* 70: 211–35.
Hájek, Alan (2019). "Interpretations of Probability." In *Stanford Encyclopedia of Philosophy*. https://plato.stanford.edu/entries/probability-interpret/
Hall, Ned (1994). "Correcting the Guide to Objective Chance." *Mind* 103 (412): 505–17.
Hall, Ned (2015). "Humean Reductionism about Laws of Nature." In Barry Loewer, and Jonathan Schaffer (eds), *Companion to David Lewis*, 262–277. New York: Wiley.
Halpin J. (2003). "Scientific Law: A Perspectival Account." *Erkenntnis* 58 (2): 137–68.
Harrison, Peter (2008). "The Development of the Concept of Laws of Nature." In Fraser Watts (ed.), *Creation: Law and Probability*, 13–36. Burlington, VT: Ashgate.
Hicks, Michael T. (2017). "Making Fit Fit." *Philosophy of Science* 84 (5): 931–43.
Hicks, Michael T. (2018). "Dynamic Humeanism." *British Journal for the Philosophy of Science* 69 (4): 983–1007.
Hicks, Michael T., and Peter van Elswyk (2015). "Humean Laws and Circular Explanation." *Philosophical Studies* 172 (2): 433–43.
Hicks, Michael T., and Jonathan Schaffer (2017). "Derivative Properties in Fundamental Laws." *British Journal for the Philosophy of Science* 68 (2): 411–50.
Hoefer, Carl (2019). *Chance in the World: A Humean Guide to Objective Chance*. Oxford: Oxford University Press.
Ioannidis, Stavros, Vassilis Livanios, and Stathis Psillos (2021). "Governing Laws and the Inference Problem." *Grazer Philosophische Studien* 98 (3): 395–41.
Ismael Jenann (2003). "Humean Disillusion." In M. Hick, S. Jaag, and C. Loew (eds), *Human Laws for Human Agents*. Oxford: Oxford University Press.
Jaag, Siegfried, and Christian Loew (2020). "Making Best Systems Best for Us." *Synthese* 197: 2525–50.
Jaynes Edwin T. (1968). "Prior Probabilities." *IEEE Transactions on Systems Science and Cybernetics* 4 (3): 227–41.
Kitcher, Philip (1989). "Explanatory Unification and the Causal Structure of the World." In Philip Kitcher and Wesley Salmon (eds), *Scientific Explanation*, 410–505. Minneapolis: University of Minnesota Press.
Kiznjak, Boris (2015). "Who Let the Demon Out? Laplace and Boscovich on Determinism." *Studies in History and Philosophy of Science Part A* 51: 42–52.
Ladyman, James, and Don Ross, with David Spurrett and John Collier (2007). *Every Thing Must Go: Metaphysics Naturalized*. Oxford: Oxford University Press.
Lam, V., and C. Wüthrich (2023). "Laws beyond Spacetime." *Synthese* 202(3): 71.

Lange, Marc (2009). *Laws and Lawmakers: Science, Metaphysics, and the Laws of Nature*. Oxford: Oxford University Press.
Lange, Marc (2013). "Grounding, Scientific Explanation, and Humean Laws." *Philosophical Studies* 164: 255–61.
Laplace, P. (1820). *Essai Philosophique sur les Probabilités* forming the introduction to his *Théorie Analytique des Probabilités*, Paris: V Courcier; repr. F.W. Truscott and F.L. Emory (trans.), *A Philosophical Essay on Probabilities*. New York: Dover, 1951.
Lazarovici, Dustin (2018). "Super-Humeanism: A Starving Ontology." *Studies in History and Philosophy of Science Part B: Studies in History and Philosophy of Modern Physics* 64: 79–86.
Lazarovici, Dustin (2020*). Typicality Reasoning in Probability, Physics, and Metaphysics*. Cham: Palgrave Macmillan.
Lewis, David (1973). *Counterfactuals*. Oxford: Blackwell Publishers and Cambridge, MA: Harvard.
Lewis, David (1983). "New Work for a Theory of Universals." *Australasian Journal of Philosophy* 61: 343–77.
Lewis, David (1986a). *Collected Papers*, vol. II. Oxford: Oxford University Press.
Lewis, David (1986b). *Philosophical Papers*, vol. II. Oxford: Oxford University Press.
Lewis, David (1994). "Humean Supervenience Debugged." *Mind* 103: 473–90.
Loar, Brian (1997). "Phenomenal States: Second Version." In Ned Block, Owen Flanagan, and Guven Guzeldere (eds), *The Nature of Consciousness: Philosophical Debates*, 597–616. Cambridge, MA: MIT Press.
Loew, Christian, and Siegfried Jaag (2020). "Humean Laws and Nested Counterfactuals." *Philosophical Quarterly* 70: 93–113.
Loewer, Barry (1996). "Humean Supervenience." *Philosophical Topics* 24(1): 101–27.
Loewer, Barry (2001a). "From Physics to Physicalism." In Carl Gillett and Barry Loewer (eds.), *Physicalism and Its Discontents*, 37–56. Cambridge: Cambridge University Press.
Loewer, Barry (2001b). "Determinism and Chance." *Studies in History and Philosophy of Science Part B: Studies in History and Philosophy of Modern Physics* 32(4): 609–20.
Loewer, Barry (2007). "Laws and Natural Properties." *Philosophical Topics* 35 (1–2): 313–28.
Loewer, Barry (2011). "Counterfactuals All the Way Down." *Metascience* 20: 27–52.
Loewer, Barry (2012). "Two Account of Laws and Time." *Philosophical Studies* 160 (1): 115–37.
Loewer, Barry (2017). "A Guide to Naturalizing Semantics." *A Companion to the Philosophy of Language*: 174–96.
Loewer, Barry (2020). "The Mentaculus Vision." In Valia Allori (ed.), *Statistical Mechanics and Scientific Explanation: Determinism, Indeterminism and Laws of Nature*, 3–29. Singapore: World Scientific.
Loewer, Barry (2021). "The Package Deal Account of Laws and Properties (PDA)." *Synthese* 199: 1065–1089.
Loewer, Barry (forthcoming a). "Typicality and Probability: Friends or Foes." In Roderich Tumulka (ed.), volume for Detlef Durr.
Loewer, Barry (forthcoming b). *The Mentaculus Vision*. In preparation.
Mach, Ernest (1960). *The Science of Mechanics*. Chicago: LaSalle Open Court.
Massimi, Micheala (2017). "Laws of Nature, Natural Properties, and the Robustly Best System." *The Monist* 100: 406–21.
Matarese, Vera (2020a). "A Challenge for Super-Humeanism: The Problem of Immanent Comparisons." *Synthese* 197: 4001–20.

Matarese, Vera (2020b). "Super-Humeanism and Physics: A Merry Relationship?" Forthcoming in *Synthese*.
Maudlin, Tim (2002). "Remarks on the Passing of Time." *Proceedings of the Aristotelian Society* 102 (3): 237–52.
Maudlin, Tim (2007). "A Modest Proposal Concerning Laws, Counterfactuals, and Explanations." In his *The Metaphysics within Physics*, 5–49. Oxford: Oxford University Press.
Maudlin, Tim (2019). *The Philosophy of Physics: Quantum Mechanics*. Princeton: Princeton University Press.
Miller, Elizabeth (2015). "Humean Scientific Explanation." *Philosophical Studies* 172: 1311–32.
Milton, John (1988). "Laws of Nature." In Daniel Garber and Michael Ayers (eds), *The Cambridge History of Seventeenth-Century Philosophy*. Cambridge: Cambridge University Press.
Ney, Alyssa (2013). "Introduction." In Alyssa Ney and David Z Albert (eds), *The Wave Function: Essays on the Metaphysics of Quantum Mechanics*, 1–51. Oxford: Oxford University Press.
North, Jill (2021). "Formulations of Classical Mechanics." In *The Routledge Companion to Philosophy of Physics*, 21–32. United Kingdom: Routledge.
Ott, Walter (2019). "Berkeley's Best System: An Alternative Approach to Laws of Nature." *Journal of Modern Philosophy* 1 (1): 4.
Ott, Walter (2022). *The Metaphysics of Laws of Nature: Rules of the Game*. Oxford: Oxford University Press.
Ott, Walter, and Lydia Patton (eds) (2018). *Laws of Nature*. Oxford: Oxford University Press.
Papineau, D. (2001). "The Rise of Physicalism." In Carl Gillett and Barry Loewer (eds), *Physicalism and Its Discontents*, 3–36. Cambridge: Cambridge University Press.
Pietroski, Paul, and Georges Rey (1995). "When Other Things Aren't Equal: Saving Ceteris Paribus Laws from Vacuity." *British Journal for the Philosophy of Science* 46: 81–110.
Popper, Karl (1959). "The Propensity Interpretation of Probability." *British Journal of the Philosophy of Science* 10: 25–42.
Psillos, Stathis (2006). "What Do Powers Do When They Are Not Manifested." *Philosophy and Phenomenological Research* 72 (1): 137–56.
Psillos, Stathis (2014). "Regularities, Natural Patterns and Laws of Nature." *Theoria* 29 (1): 9–27.
Psillos, Stathis (2018). "Laws and Powers in the *Frame of Nature*." In Walter Ott and Lydia Patton (eds), *Laws of Nature*, 101–124. Oxford: Oxford University Press.
Quine, Willard Van Ornam (1981). *Theories and Things*. Cambridge, MA: Harvard University Press.
Roberts, John T. (2008). *The Law-governed Universe*. Oxford, MS: Oxford University Press.
Ruby, Jane (1986). "The Origins of Scientific Law." *Journal of the History of Ideas* 47: 341–59.
Schaffer, Jonathan (2007). "Deterministic Chances?" *British Journal for the Philosophy of Science* 58 (2): 113–40.
Schaffer, Jonathan (2010). "Monism: The Priority of the Whole." *Philosophical Review* 119 (1): 31–76
Schaffer, Jonathan (2016). "It Is the Business of Laws to Govern." *dialectica* 70 (4): 577–88.

Schaffer, Jonathan (2017). "Laws for Metaphysical Explanation." *Philosophical Issues* 27 (1): 302–21.
Schrenk, Markus (2007). *The Metaphysics of Ceteris Paribus Laws*. Frankfurt: Ontos.
Shafer, Glenn (2006). "Why Did Cournot's Principle Disappear." http://glennshafer.com/assets/downloads/disappear.pdf.
Shoemaker, Sydney (1980). "Causality and Properties." In Peter van Inwagen (ed.), *Time and Cause*. Dordrecht: D. Reidel Publishing Company.
Strawson, Galen (1989). *The Secret Connexion: Causation, Realism, and David Hume*. Oxford: Oxford University Press.
Tooley, Michael (1977). "The Nature of Laws." *Canadian Journal of Philosophy* 7: 667–98.
Van Fraassen, Bas (1989). *Laws and Symmetry*. Oxford: Oxford University Press.
Van Lunteren, Frans (2016). "The Missing History of the Laws of Nature." http://www.shellsandpebbles.com/2016/11/07/the-missing-history-of-the-laws-of-nature/
Verhaegh, Sander (2018). *Working from Within: The Nature and Development of Quine's Naturalism*. Oxford: Oxford University Press.
Weinberg, Steven (2011). *Dreams of a Final Theory: The Scientist's Search for the Ultimate Laws of Nature*. New York: Knopf Doubleday Publishing Group.
Wilhelm, Isaac (2002). "Typical: A Theory of Typicality and Typicality Explanation." *The British Journal for the Philosophy of Science* 73 (1): 1–19.
Williams, D. C. (1951). "The Myth of Passage." *The Journal of Philosophy* 48 (15): 457–72.
Williamson, Jon (2010). *In Defense of Objective Bayesianism*. Oxford: Oxford University Press.
Woodward, James (2018). "Laws: An Invariance Based Account." In Walter Ott and Lydia Patton (eds), *Laws of Nature*, 158–180. Oxford: Oxford University Press.
Wüthrich, Christian, and Vincent Lam (2023). "Laws beyond Spacetime." *Synthese* 202 (3): 1–24.

Index

For the benefit of digital users, indexed terms that span two pages (e.g., 52–53) may, on occasion, appear on only one of those pages.

Albert, David vii, 13 n.16, 20 n.5, 26–7, 47 n.27, 86
Aristotle 1–2
Aristotelian conception of science 1–2
Armstrong, David 9–11, 13–15, 38–9, 53–4, 54 n.2, 67, 104

Balog, Katalin 84 n.11
Barbour, Julian 80
Bell, John 80, 80 n.2
Belot, Gordon 126 n.2, 128
Best systems account of laws and chances (BSA) 5–8, 18–54, 43 n.22, 56 n.7, 57, 59–61, 64 n.24, 64–5, 68–9, 71–3, 72 n.3, 76–81, 78 n.10, 87–90, 92, 94–5, 101 n.9, 105–6, 108–15, 122–9, 132–4, 136
Bhogal, Harjit 43 n.22, 58 n.12, 138
Bird, Alexander 14
Blanchard, Thomas 27 n.18, 43 n.22

Callander, Craig 117–18
Carnap Rudolf 100–1
Carroll, Sean 12, 93–4
Chen, Eddy 12–14, 30, 38–9, 58 n.13
Coffey, Kevin 62 n.21
circularity objection to Humean accounts 40–3
Clarke, Samuel 2
Cohen, Jonathan 117–18
Counterfactuals 3, 5, 10–11, 14, 16, 18–19, 22–4, 33, 35–40, 42–3, 77 n.9, 77, 94, 113–14, 120, 136–7
  Cotenability 35, 35 n.8

Cournot's Principle 98, 104–5
Cotenability 35
  Probabilistic accounts 111–12
  Similarity accounts 35
Cohen, Jonathan 118

Demarest, Heather 66
Dasgupta, Shamik 67–70, 124
De Finetti, Bruno 100
Descartes, Rene 1–3, 25, 117–18
Determinism 96–8, 101–2, 106
  Boscovich 3
  Laplace 3–5
Dorr, Cian 69 n.31
Dorst, Chris 23, 38
Dretske, Fred 9–14

Eagle, Anthony 105
Earman, John 4 n.13
Einstein, Albert 24, 80–1
Elga, Adam 112–13
epistemological holism 116
Esfeld, Michael 12 n.14, 64, 71–80
Euthyphro problem 70

Fermat, Pierre 5
Field, Hartry 43 n.23
Filomeno, Aldo 45
Fodor, Jerry 116
Foster, John 4, 45–7
Friedman, Michael 43 n.22
Frigg, Roman 113–14
Fuchs, Christopher A 99 n.5

Galileo 1–3
God 2–4, 51
　Creator of the laws 2–3, 7–8
　God's eye view 95
Goldstein, Shelly 12 n.14, 13, 103 n.12
Goodman, Nelson 19 nn.2, 3, 68–9, 129–30
Gomez, Veronica 57 n.8
governing laws 3–4, 7–13, 15, 33, 39–40, 44, 103 n.12
Guo, Bixin 124 n.1

Hacking, Ian 5–6, 98
Hajek, Alan 96 n.1, 102 n.11, 103 n.12
Hall, Ned 23, 32, 108–9
Halpin James 34 n.7
Harrison, Peter 1–2
Hawking, Stephen 1, 3, 6, 121
Hicks, Mike 27 n.17, 41–2, 61–2, 113–14, 120 n.8
Hoefer, Carl 105 n.17
Hume, David 6, 8, 18–19, 40, 43, 96
Humean mosaic (HM) 8 n.5, 21–2, 40–1, 98, 110–12
Humean supervenience (HS) 21–2, 55, 58–9

inference problem 9–10, 104
Ioannidis, S. 9 n.9
Ismael, Jenann 44, 49–50

Jaag, Siegfried 23, 27 n.17
Janus-faced 5, 98
Jaynes, E.T. 100

Kitcher, Philip 43 n.22
Kiznjak, Boris 3 n.7

Ladyman James 54 n.2, 64 n.22
Lange, Marc 8 n.4, 24 n.14, 41–2
Laplace, Simon 3–4, 96–8
Laws of nature
　Best systems account (BSA) 5–8, 20–5, 27, 29–35, 39–50, 53–4, 56–7, 61–2, 66–81, 88–90, 94–5, 105–6, 110–12, 114–15, 118, 124–6, 128–37

Better best systems account 118
Constraint Accounts 12–13
Ceteris paribus laws 6, 118–19
Dynamical laws 10–12, 23, 38–9, 106–7
Explanation 10–12, 38–50, 106–7, 120, 135–6
FLOTES 10–12
Governing 3–4, 7–13, 15, 33, 39–40, 44, 103 n.12
Humean 5–8, 14, 18–22, 27, 29, 33, 40–50, 53–4, 56–7, 61–2, 66–81, 88–90, 94–5, 105–6, 110–12, 114–15, 118, 124–6, 128–37
induction 43–5, 49–50, 69, 136
Non-Humean 5–13, 15, 18–19, 27, 29, 33–4, 39–40, 43–5, 49–50, 69, 77–8, 99, 104–5, 114, 132–3, 135–6
Package Deal Account (PDA) 5–6, 12 n.14, 49–50, 54–5, 57–8, 60–1, 64–5, 67–70, 76–96, 110–11, 113, 117–37
Powers Account 8, 12, 14–15, 48, 76–7, 89
Special science laws 6, 116–23
Systematizing and unifying 4–5, 7–8, 13, 18–21, 24–5, 29, 42–3, 74–7, 79–83, 88–9, 94–5, 105, 117–23, 132–3, 135–6
Thought experiments 33–4
Lazarovici, Dustin 47–8, 78
Lewis, David 5–6, 8–9, 20–2, 26–7, 32, 36–7, 39–40, 53–7, 61–2, 66, 71–2, 79–80, 96, 104–9, 112–13, 126–7, 130–2
Livanios, V. 9–10
Loar, Brian 84–5
Loew, Christian 23, 118

Mach, Ernest 80
Massimi, Micheala 32
Matarese, Vera 74–5
Mentaculus 4–5, 93 n.16, 110 n.25, 120, 122

Mentaculus Diagram Probability Map of the Universe 121
Millikan, Ruth 130–1
Milton, John 4
Missing value problem 67–9, 135–6
Mismatch problem 61–2, 66, 134

Natural properties
　Degrees of naturalness 69
　Perfectly natural properties 5–6, 8, 20–2, 53–7, 61–2, 66, 71–2, 79–80, 110 n.25, 130–2
　Problems with perfectly natural properties 53–70
　Categorical or dispositional 8, 54, 59–60, 63–4, 74, 88–9, 134–5
　mismatch problem 61–2, 66, 134
　Missing value problem 67–9, 135–6
　Euthyphro problem 70
naturalizing metaphysics 6, 57, 79
necessary connections 7–8, 14, 18–19, 39–40, 88–9
Newton, Isaac 1–5, 9–14, 35, 57–8, 71, 81, 116
Ney, Alyssa 58–9, 94–5
North, Jill 62–3

Oppenheim, Paul 116
Ott, Walter 7–8

Package Deal Account of Laws, Chances, Space-time and Fundamental Properties (PDA) 5–6, 29, 69, 74–7, 79–95, 114, 117–20, 124–6, 128–9, 132–9
Papineau, David 119–20
Pascal, Blaise 5, 96
Past hypothesis (PH) 13, 30–2, 92–3, 121, 134–5
Perry, Zee 58–9, 73–4
Physicalism 4–6, 119–20
　Causal closure argument 119–20
Pietroski, Paul 118–19
Poisson, Siméon Denis 96, 98
Popper, Karl 99, 103

Possible worlds 18–19, 36–7, 107–8
　Account of counterfactuals 36–7
　Account of necessity 18–19
probability/chance 5–6, 96–115, 132
　actual frequency 102–3
　best systems account 20–1, 105–6, 110–11
　Cournots principle 99
　epistemic probability 97–8, 109–10
　hypothetic frequency account 102–3
　law of large numbers 96–7
　Principle of indifference 100–1
　Principal Principle (PP) 99, 108–9, 111–12
　propensity account 8, 103–4
　randomness 46–7
Psillos, Stathis 4–5, 14–15
Putnam, Hilary 54, 129–31
Putnam's Paradox 66, 129–31

Quantum mechanics
　Bohmian mechanics 12, 58–9, 109
　QBism 73
Quine, W.V.O. 6, 81, 139

Ratbag idealism 26–7, 68–9, 126–9
Realism 6, 124–31
　Metaphysical realism 53, 57, 124–31
　Putnam\u2019s Model theoretic argument 129–31
　Scientific realism 125–6, 129–30
Reference 6, 54, 66, 86, 91, 130–1
Reference magnetism 66, 130–1
Relativism 6, 124, 128, 131
Rey, Georges 118–19
Roberts, John 31–2
Ross, Dom 57–8
Ruby, Jane 1–2
Russell, Bertrand 64, 96

Scientifically fundamental properties 87–9, 91, 132
Schaffer, Jonathan 10–11, 82, 97–8, 107, 118–19
Shafer, Glenn 99
Shoemaker, Sidney 15

INDEX

Schrenk, Markus 30, 118
Schrödinger's equation 10, 14
Sider, Ted 61
space-time 4, 11–13, 20, 28, 64, 71–2, 80–1, 134
Special science laws 6, 30, 79–80, 116–20, 122–3, 132–3
Standard model 59–60
Strawson, Galen 1–2
super Humeanism 71

thermodynamics 2, 4–5, 119–20
Theory of everything (TOE) 2–3, 6, 61–2, 81–2, 117
Time 4–5, 11–12, 96–7, 121
  arrows of 11–12, 96–7, 121
  block universe 28, 114–15
  direction 11–12
  growing block 28
  presentism 28
Tooley, Michael 9
typicality 47–8

van Fraassen, Bas 10, 13, 61, 66, 98
van Lunteren. Frans 7
Verhaegh, Sander 81, 139

Weinberg, Steven 2–3, 61–2, 81
Wilhelm, Isaac 47–8
Williams, D.C. 134–5
Williamson, Jon 100
Woodward, James 23
Wüthrich, Chris 12–13, 80–1